High Performance Liquid Chromatography - Recent Advances and Applications

Edited by Oscar Núñez, Sònia Sentellas, Mercè Granados and Javier Saurina

Published in London, United Kingdom

High Performance Liquid Chromatography - Recent Advances and Applications
http://dx.doi.org/10.5772/intechopen.105269
Edited by Oscar Núñez, Sònia Sentellas, Mercè Granados and Javier Saurina

Contributors
Ayenew Ashenef, Ariaya Hymete, Tadele Eticha, Feruza Ahmed, Natalija Nakov, Jelena Acevska, Katerina Brezovska, Zoran Kavrakovski, Aneta Dimitrovska, Francesca Di Gaudio, Sergio Indelicato, Annamaria Cucina, Mohamed Hefnawy, Manal El-Gendy, Ali El Gamal, Prafulla Kumar Sahu, Sukumar Purohit, Swarnajeet Tripathy, Durga Prasad Mishra, Biswajeet Acharya, Magali Monteiro, Maria Rita Estevam, Estela Mesquita, Kelly da Silva Bezerra, Seyit Yuzuak, Qing Ma, Yin Lu, De-Yu Xie

First published in London, United Kingdom, 2023 by IntechOpen
IntechOpen is the global imprint of INTECHOPEN LIMITED, registered in England and Wales, registration number: 11086078, 5 Princes Gate Court, London, SW7 2QJ, United Kingdom

British Library Cataloguing-in-Publication Data
A catalogue record for this book is available from the British Library

Additional hard and PDF copies can be obtained from orders@intechopen.com

High Performance Liquid Chromatography - Recent Advances and Applications
Edited by Oscar Núñez, Sònia Sentellas, Mercè Granados and Javier Saurina
p. cm.
Print ISBN 978-1-83768-565-3
Online ISBN 978-1-83768-566-0
eBook (PDF) ISBN 978-1-83768-567-7

Meet the editors

Oscar Núñez studied chemistry at the University of Barcelona, where he also obtained his Ph.D. in 2004. Since June 2020 he has been a professor in the Analytical Chemistry Department of the same university. He has more than 150 scientific papers and book chapters to his name, and he is the editor of seven books on liquid chromatography, LC-MS/MS, sample preparation techniques in food analysis, capillary electrophoresis, and food integrity and authenticity. He has extensive experience in the development of liquid chromatography methods with ultraviolet and fluorescence detection, liquid chromatography coupled to low- and high-resolution mass spectrometry, as well as in sample treatment procedures for environmental and food analysis. Currently, his main research interests involve the characterization, classification, and authentication of food and natural products, as well as the prevention of food fraud.

Sònia Sentellas has been a Serra Húnter lecturer in the Department of Chemical Engineering and Analytical Chemistry of the University of Barcelona since 2020. She had previously worked as a researcher in the pharmaceutical company Almirall, S.A. for nearly 20 years, focusing on the study of the metabolism of drugs and bioanalysis. Her research interests cover food analysis, including characterization, classification, and authentication of food products, fraud prevention, and the recovery of bioactive compounds from agrifood waste. She is a member of the Research Institute for Food Nutrition and Food Safety.

Mercè Granados, Ph.D., is an Associate Professor in the Department of Chemical Engineering and Analytical Chemistry, at the University of Barcelona. She has supervised master's and Ph.D. students and has published more than 80 scientific papers. Her wide research interests include topics related to environmental analytical chemistry and the fate of contaminants, as well as the analysis of residues in food, with liquid chromatography and mass spectrometry being the main techniques applied in these studies. Currently, she is involved in circular-economy projects related to the recovery of bioactive substances from agrifood waste.

Javier Saurina is a full professor in the Department of Chemical Engineering and Analytical Chemistry at the University of Barcelona. A graduate in chemistry (1988) and pharmacy (1996), and with a Ph.D. in chemistry (1997), his current research focuses on the development of new analytical methods, the characterization and authentication of food products, and data processing. He has published more than 20 book chapters and 190 articles in indexed journals in the fields of analytics, pharmaceuticals, materials science, and food. He is among the 2% most cited and influential authors according to the updated science-wide author databases of standardized citation indicators published in August 2021 (see https://elsevier.digitalcommonsdata.com/datasets/btchxktzyw/3).

Contents

Preface

Increasing interest in topics related to health and quality of life in recent years has led to a growing need in food, environmental and bioanalytical research for high-throughput separation techniques able to cope with the qualitative/quantitative determination of a large number of compounds in very complex matrices. High-performance liquid chromatography (HPLC) is a well-established separation technique widely employed in many fields. The versatility of chromatographic separation modes, coupled with low-resolution and high-resolution mass spectrometry, makes HPLC among the best options to solve emerging analytical problems.

This book provides an overview of new advances in high-performance liquid chromatography and its applications in different fields. The first of the book's two sections is devoted to principles and new advances in HPLC. In Chapter 1, "Principles and Applications of Ultra-High-Performance Liquid Chromatography", Ahmed et al. describe the fundamentals of UHPLC techniques (solvent delivery systems, columns, sample injection, and detection), and review the role of UHPLC in different fields such as the analysis of herbal medicines, drugs in human plasma, pharmacokinetics and bioavailability studies, identification of metabolites (metabolomics), detection of impurities, analysis of dosage formulations, food safety and agricultural applications.

The higher separation capacity of 2D-LC techniques can be especially useful. A commercial application, the TurboFlow technology developed by Thermo Fisher Scientific, may enable the direct introduction of biological samples into an online automated extraction system without any pre-treatment. In Chapter 2, "Turbulent Flow Chromatography: A Unique Two-Dimensional Liquid Chromatography", Di Gaudio et al. explain the principles of turbulent flow chromatography, the different column chemistries available, the hardware employed, how to operate in multichannel systems, and how to develop a focus mode method to implement this technique. They also describe successful applications of the turbulent flow technique in different fields, including therapeutic drug monitoring and environmental analysis, and applied to different matrices such as urine, plasma, food commodities and water.

Monolithic media are excellent substitutes for conventional particle-packed columns. This is because monolithic columns show higher permeability and lower flow resistance than conventional liquid chromatography columns, providing high-throughput performance, resolution and separation in short run times. In Chapter 3, "Monoliths Media: Stationary Phases and Nanoparticles", Hernawy et al. address the general characteristics and properties of monolithic materials, the separation mechanism involved, and the different types of monoliths: organic polymer, silica and metal. Applications of monolithic materials in LC separations and nanoparticle-based monoliths are also described.

In Chapter 4, "Perspective Chapter: High-performance Liquid Chromatography Coupled to Mass Spectrometry – The Advance in Chemical Analysis", da Silva Bezerra presents the advantages and limitations of LC-MS techniques, addressing the principles of coupling, sample preparation and other analytical issues and describing relevant LC-MS(/MS) applications in different fields.

The second section is devoted to applications. Chapter 5 by Nakov et al., "Green Strategies toward Eco-Friendly HPLC Methods in Pharma Analysis", describes green chemistry strategies that can be easily applied to conventional liquid chromatography instruments to develop eco-friendly HPLC methods for pharma analysis. Several tools for the evaluation of the greenness of chromatographic methods are also described.

The next two chapters are devoted to the application of HPLC techniques in the identification and determination of polyphenolic compounds, important bioactive substances with antioxidant properties, in food products. Chapter 6, "Phenolic Compounds Profile of Brazilian Commercial Orange Juice", by Estevam et al., addresses the characterization of bioactive compounds such as flavonoids and phenolic acids in orange juice from Brazil using HPLC-DAD and UHPLC-ESI-MS, together with multivariate analysis. In Chapter 7, "HPLC-MS$^{(n)}$ Applications in the Analysis of Anthocyanins in Fruits", Yuzuak et al. review the application of HPLC-MS/MS in the identification and characterization of anthocyanins. These compounds are water-soluble pigments found abundantly in the flesh and skin of fruits, flowers, and roots of different varieties of plants. Correct identification of anthocyanins with similar structures and accurate estimation of their contents within fruit matrices remain challenging. LC-MS/MS-based metabolomic studies have been found to be an effective technology for distinguishing anthocyanins that are similarly structured.

In Chapter 8, "Current Trends in HPLC for Quality Control of Spices", Kumar Sahu et al. address the role of LC in the characterization and quality control of spices and condiments that have important antioxidant, antibacterial, anticancer and anti-inflammatory properties. Both aflatoxins secreted after fungal contamination and illegal dyes used to adulterate spices cause quality degradation. HPLC is a rapid and adaptable technique for the quality assessment of spices.

We hope this book will be useful to all in the scientific community dealing with the development of HPLC analytical methods and their applications in different fields.

Oscar Núñez, Sònia Sentellas, Mercè Granados and Javier Saurina
Department of Chemical Engineering and Analytical Chemistry,
University of Barcelona,
Barcelona, Spain

Section 1

Principles and New Advances

Chapter 1

Principles and Applications of Ultra-High-Performance Liquid Chromatography

Feruza Ahmed, Tadele Eticha, Ariaya Hymete and Ayenew Ashenef

Abstract

The science of separation had advanced significantly with the development of ultra-high-performance liquid chromatography (UHPLC), a brand-new type of liquid chromatography. The need for the evolution of HPLC into UHPLC has been driven by the continuously evolving of packing material modifications that affect the separation of mixtures. The separation process of analytes is completed in a substantially decreased amount of time due to the lower particle sizes, which increases surface area of interaction allowing reduction of column length to one-third; thus, shorter columns are employed in UHPLC, which consequently causes the flow rate to be three times higher and subsequently reducing analysis time. Although UHPLC shares the same fundamental idea and instrument layout as HPLC, it differs from HPLC in that it produces narrow peaks and has high spectral quality, allowing for simple compound identification in a variety of analytical applications such as impurity profiling, product formulation, and improved analytical technique and method development. However, high back pressure in UHPLC might lead to decreased column life, and the instrument's higher price compared to HPLC are the disadvantages.

Keywords: high-performance liquid chromatography (HPLC), particle size, ultra-high-performance liquid chromatography (UHPLC), principles of chromatography, applications of chromatography

1. Introduction

Chromatography is a separation method that distributes a mixture's components between a stationary phase and a mobile phase via several methods, including adsorption, partition, ion exchange, and others [1]. The most popular method to identify, measure, and separate the components in a mixture is liquid chromatography (LC). Later, LC developed into high-performance liquid chromatography (HPLC), which pushes solvents through a column under high pressure [2]. It is an effective LC method for separating mixtures. Additionally, it is utilized to identify and measure pharmaceuticals in biological fluids, final dosage forms, and during the drug research and discovery process. Instrumental developments have been made and are still being made to improve resolution and other separation-related properties such as speed and sensitivity [3].

The development of UHPLC is a result of the growing need for quick and ultra-quick separation techniques that are more effective and have superior resolution [4]. UHPLC has ushered in a substantial shift by giving analysts new ways to get quick analytical separation techniques without compromising the high-quality outcomes previously attained by HPLC [5].

UHPLC has an astonishingly short analysis time and uses a very small amount of solvent as the mobile phase. Additionally, it significantly increases separation effectiveness and analyte mixture resolution. UHPLC uses column packing particle size of less than 2 microns as its key differentiator from conventional HPLC systems, which use particles between 2.5 and 10 microns in size. Because the smaller particles (2 microns) require a greater pressure to work with, UHPLC systems must be able to perform over 6000 psi, which is frequently the upper limit of conventional HPLCs [5, 6].

2. Principles of ultra-high-performance liquid chromatography

The first UHPLC systems appeared in 2004 [7]. The fundamental idea behind this modification of HPLC is that efficiency is gained as column packing particle size lowers, a particle size reduction of less than 2 μm results in an improvement in efficiency that does not drop at higher linear velocities or flow rates [3, 5].

By boosting chromatographic resolution with the greatest number of resolvable peaks, UHPLC enhances the separation systems of LC. By using small amount of column packing materials and reduction of particle size, analysis becomes faster and more sensitive. Additionally, UHPLC has enhanced instrument designs [8]. It had a shorter analysis time of about 1.5 minutes and it reduces the mobile phase volume usage by at least 80% when compared to HPLC. In general, the development of UHPLC has brought a significant advantage for analysts by providing quick and accurate analytical separation findings [5].

3. Distinct instrumental designs of ultra-high-performance liquid chromatography

The common components of UHPLC system are solvent delivery systems (Pumps), sample injection, columns, column managers, detectors [3].

3.1 Solvent delivery system

UHPLC systems regularly operate at 8000–15000 psi. The delivery system must also counterbalance for various solvents used in isocratic, linear and nonlinear gradient elution modes, and also for solvent compressibility for a wide range of pressures. The two major classifications of solvent delivery systems are constant pressure pump and constant volume pump [2, 9, 10]. Constant pressure is used for column packing while constant volume pump is mostly used in all common UHPLC applications. HPLC has a pump pressure of 40 Mpa, whereas the UHPLC has a pump pressure of 100 Mpa [7].

3.2 UHPLC columns

UHPLC columns are short in length and have a 150 × 2.1 mm length and diameter dimension respectively with a smaller diameter that ranges from 1 to 2.2 mm [10, 11]. Capillary columns are particularly suitable for UHPLC systems due to lower heat generation and better heat tolerance capability. These columns can operate with pressures higher than 80,000 psi [2]. Charged surface hybrid particle technology, ethylene bridged hybrid particle technology [1, 5, 6], high strength silica particle technology, and peptide separation technology are most commonly used in the construction of columns used in UHPLC [7, 10, 12, 13].

3.3 Column manager

The Column manager adjusts temperature from 10 to 90°C and switches automatically for up to 24 hours to four columns, each with a diameter of 2.1 mm internal diameter and length of 150 mm. It also has the ability to bypass channel for flow injections. Most UHPLC systems contain a binary solvent manager, sample manager with the column heater, detector, and non-compulsory sample organizer. The binary solvent manager employs two individual serial flow pumps to deliver a parallel binary gradient. There is a built-in solvent selection feature valves that allow one to pick among the availed up to four different solvents [14].

3.4 Sample injection

The volume of the sample in UHPLC is usually 2–5 μl. Injection cycle time is 25 s without a wash and 60 s with a dual wash used to further decrease carry-overs. A variety of microtiter plate formats (deep well, mid height, or vials) can also be accommodated in a thermostatically controlled environment when analyte stability and conditions demand. Using the optional sample organizer, the sample manager can perform injection from up to 22 microtiter plates [3]. There is also a direct injection application for biological samples [15].

3.5 The detector

The most common detectors used in UHPLC analysis are UV/visible-based types. Detection of analytes is conventionally based on absorbance [9]. The UHPLC detector ought to have long path length, low volume detection cell, a highest likely sensitive detection, and reliable quantification of the narrow peaks. System volumes should also be lessened to uphold the speed, resolution, and sensitivity of the analysis [10]. Depending on the type of detector, the sensitivity of UHPLC can be increased by 2–3 times more than that of HPLC [2].

4. Applications of UHPLC

UHPLC is playing a substantial role in the advancements of liquid chromatography. This is highly attributed to its ability in providing efficient and fast analysis. It can also be hyphenated with different instruments that make application in immense territories like that of pharmaceutical, toxicological, and food industry. It is helping to determine the nutritional value of certain types of foods. It had wide applications in different agricultural sectors and in clinical analysis, where it is vital to increase throughput with reduced analysis costs [16].

4.1 Analysis of herbal medicines

Herbal medicines are not pure products with a single active ingredient. Thus, conventional methods for screening and identifying the active ingredients in natural products are inefficient. Traditional natural product discovery, using conventional methods, does not give enough evidence about mode of action until late stage in the discovery course. This causes finding compounds with exceptional biological properties (single compound/API drug discovery) a difficult task [17].

Although herbal medicines are gaining increasing attention, clinical usage studies that evaluate its safety and efficacy based on ***in vivo*** pharmacokinetic data of its main ingredients are limited [18].

Chromatographic fingerprinting of herbal components by UHPLC–MS has become a powerful and widely used technique today. This is because it could systematically profile the composition of herbal medicine samples. It also provides high-quality separations and detection capabilities for active compounds in highly complex samples derived from natural origin herbal remedies [19–21]. The application of UHPLC-HRMS for quality control of traditional Chinese medicine (TCM) includes chemical characterization of TCM, determination of TCM components, chemical fingerprint analysis, identification of the authenticity of TCMs, identification of illegal additives in TCMs, exploring the Quality-Marker (Q-Marker), identification of metabolites, evaluation of the quality of TCMs from different habitats, elucidation of the mechanism of action of TCMs (10).

Table 1 shows typical applications of UHPLC in the analysis of herbal medicines where different methods of UHPLC were employed.

The pulp, seeds, and peel of the seven tropical fruits were tested, and the antioxidant levels and capacities revealed notable variations. Fruits' peels and seeds contain more antioxidant potential and antioxidant chemicals than the pulp, which is deficient in these substances. Most of the samples' antioxidant component levels are reduced by oven and freeze drying process. The results of this study had shown how valuable sources of natural antioxidants in avocado-dried peel and passion fruit seed could be affected by the processing techniques. By using UHPLC-ESI-MS, organic acids (citric, malic, and tartaric) and phenolic compounds were also quantified [17].

By using UHPLC/QTOF MS together with automated identification by the Metabolynx^{TM} instrument, a total of 33 peaks were tentatively identified ***in vitro***, 24 of which were parent components, with nine metabolites being detected: This offered beneficial chemical data for additional pharmacological investigation [22].

The validated UHPLC-ESI-MS/MS method was used in the PK study of 14 ingredients after oral administration of Gumiganghwal-tang tablets, a TCM [18].

More importantly, the application of the UPHLC/MS-network pharmacology method will offer a more trustworthy and convincing tool to identify future prospective targets and biological processing mechanisms of Chinese medicine [19].

In an investigation, acetylcholinesterase inhibition assay ***in vitro*** were utilized to quickly screen and discover acetylcholinesterase inhibitors in the D. auriculatum using UHPLC CQ-TOF-MS and UHPLC-ESI-MS/MS. The technique could be utilized for the quick discovery of novel AChEI from natural compounds and was straightforward, sensitive, and selective. This study had provided scientific experimental basis for the traditional efficacy of the medicinal plant for neurological disease [20].

DOI: http://dx.doi.org/10.5772/intechopen.110540

Experimental Condition	Application	Analytes	Matrix	Type of column	Oven T°	Flow rate	MP	Finding	Ref.
UHPLC–ESI–MS	Antioxidant property	Avocado pineapple, banana, papaya, passion fruit, watermelon and melon	Fruit extracts	*	30°C.	0.20 mL/ min.	*	Natural antioxidants	[17]
UHPLC/QTOF MS	Therapeutic effect of ShenQiWan (SQW) treatment of kidney-yang deficiency syndrome	ShenQiWan (SQW)	Serum	Λ	40°C	0.5 ml/min	Λ	Therapeutic effect on kidney-yang deficiency syndrome	[22]
UHPLC-Q Orbitrap MS	Metabolite profiling and pharmacology	XIAOPI formula extract samples	Rat serum	%		0.5 mL/min	%	Potential critical role of bile acid synthetic pathway	[19]
UHPLC-MS	AchE inhibitory activity characterizing of phytochemical constituents	*D. auriculatum*	Extracts of genus Dichocarpum and other 6 species	#	35°C	0.3 mL/min	#	Significant AChE inhibitory activity	[20]
UHPLC-MS/MS	Herb-Drug interaction	*Polygonum capitatum* extract and Levofloxacin (LVFX).	Urine and fecal matter of rat	$	40°C	0.3 mL/min	$	Drug interaction exists	[21]
UHPLC-Q-Extractive-Orbitrap MS	Interventionl effects	Sanse powder	Rat serum	@	6°C	0.3 mL/ min	@	Protective effects against knee osteoarthritis in rats	[23]

*UHPLC column: *UHPLC BEH C18 1.7 μm particles (2.1 × 50 mm) column, Λ BEH C18 column (100 mm × 2.1 mm i.d., 1.7 μm particle size, %UHPLC HSS T3 column (2.1 mm×100 mm, 1.8 μm, Waters), #C18 column (1.6 μm, 2.1 mm×100 mm, $BEH C18 Column (2.1 mm × 100 mm, internal diameter 1.7 μm), @CSH C18 column (1.7 μm, 2.1 × 100 mm), Mobile Phase: * (Mp-A) formic acid0.1% in Milli-Q water and (MP-B) methanol. Λ (Mp-A) HCOOH to H2O ratio of 1:1000, v/v and (Mp-B) HCOOH to CH3CN of 1:1000, v/v. %(MP-A) 0.1. % formic acid in water and 5 mmol/L ammonium acetate in water (MP-B) was acetonitrile. # (Mp-A) acetonitrile and (Mp-B) water, contains 0.1% formic acid. $ 0.1% aqueous formic acid (Mp-A) and (Mp-B) 0.1% formic acid in acetonitrile @(Mp-A)6:4 acetonitrile: water + 10 mM ammonium formate + 0.1% formic acid and (MP-B) 9:1 isopropanol: acetonitrile + 10 mM ammonium formate + 0.1% formic acid.*

Table 1.
Application of UHPLC in natural and herbal medicines analysis.

UHPLC has been employed to study a herb-drug interaction in animal experiments, that is, rats. It was used to evaluate the tissue distribution and excretion of ***Polygonum capitatum*** extract and Levofloxacin when co-administered on rats [21].

4.2 Analysis of drugs in human plasma

Detection of drugs in biological samples is a requisite to study the pharmacokinetics, toxicity, and bioequivalence of the medicine [5]. The most crucial part of determination of drugs in biological samples (plasma, serum, urine, saliva, etc.) is sample preparation. It is often a baseline for a reliable and fast approach in refining analytical efficiency. Blood sample contains numerous proteins and other possible interfering components that would affect the detection of analytes unless otherwise eliminated. Protein precipitation, solid-phase extraction (SPE), and liquid–liquid extraction (LLE) have been widely employed for sample preparation to recover sufficient analyte from the biological sample matrices [24].

For example, a method was developed and validated for drug adherence measurement of patients with hypertension. The antihypertensive drugs were analyzed from patients' blood using a simple and fast sample preparation protocol with protein precipitation followed by chromatographic separation using a gradient elution on a reversed phase column. Mass spectrometric detection was conducted by applying both positive and negative electrospray ionization (ESI+/ESI−) and selected reaction monitoring mode (MS/MS). Only 50 μl of plasma sample was needed for the simultaneous quantification of all twelve compounds (Amlodipine besylate, hydrochlorothiazide, nifedipine, spironolactone, valsartan, canrenone, enalapril, losartan, losartan carboxylic acid, perindopril and perindoprilate, enalaprilat) within 6-min runtime. Enalapril-d5 was applied as internal standard for all compounds except hydrochlorothiazide in which case the internal standard was hydrochlorothiazide-$^{13}C,d_2$. The method showed a significant advantages of minimal sample volume, clean-up procedure, and a short runtime. The method is now available to monitor drug adherence of patients thus helping to manage resistant hypertension in a hospital setting [25].

Analysis of drugs in human plasma using UHPLC plays another significant role in the therapeutic drug monitoring (TDM) activities. The method is quite simple and rapid (separation of compounds achieved in only 6 min) using just a precipitation method for sample preparation and utilizing low amount of patient plasma (100 μL). Due to these characteristics, the UHPLC-based method allows processing of clinical samples and rendering of results within the same working day. This makes it advantageous and fit for TDM purposes [26].

UHPLC has also been employed in the forensic sector. It analyzes compounds of forensic interest in human plasma. A study was carried out on Diphenidol (DPN), a non-phenothiazinic antiemetic, and antivertigo drug following some cases of suicide and accidental poisoning in China. Even though the exact mechanism of death caused by DPN poisoning has not been fully explained yet it is thought that an excessive amount of DPN may promote successive H1 receptor antagonist and anticholinergic effects, causing depression in the central nervous system and vascular smooth muscle relaxation, resulting in sedation, lethargy, hypotension, convulsions, and respiratory failure. The study had been performed using UHPLC–MS–MS indicated that the postmortem concentration range of heart blood was $0.87 \sim 99$ μg/mL. The method was successfully applied to the detection and quantification of DPN in 15 real forensic cases [27]. **Table 2** lists UHPLC's deployment in drug analysis from human plasma.

Traditional methods are typically time-consuming, non-quantitative, and complex, and the necessary sample size is usually big, making them challenging. To fulfill the requirements of quick clinical detection and forensic identification, it is crucial to design simple, sensitive, and accurate systems. UHPLC analytical technique was effectively utilized for quantitative analysis of diphenidol in tissues where blood samples are unavailable or of poor quality [27].

In comparison with previously published methods, the recently developed UHPLC approach is quick and inexpensive, and has superior selectivity and sensitivity for the simultaneous quantification and pharmacokinetics study of human medications in their pharmaceutical dose forms and in spiked human plasma [28–30].

4.3 Pharmacokinetics and bioavailability study

UHPLC method has been developed and validated for the quantitative measurement and determination of pharmacokinetics and bioavailability of various drugs. This had a significant task in assuring drugs' quality, safety, and efficacy. Besides, since most preclinical studies are done on animals, there has been an indication of potentially massive savings of the number of animals and the amount of compounds used when efficient UHPLC methods are employed [31].

UHPLC is also useful for the study of pharmacokinetics and bioavailability on herbal medicines. UHPLC–MS/MS method was developed for the determination of Uncaria alkaloids (a TCM of dry hooked branch of Rubiaceae) which has shown an antihypertensive, vasodilating, neuroprotective, antidepressant, antiarrhythmic, antiepileptic, and antitumor activities in the blood of mice. The relevant methodological process was verified using UHPLC analytical methods. The results showed that the established UHPLC–MS/MS method is accurate and fast. It takes only 5.5 min to analyze blood sample. It is highly sensitive, and effective for the detection and pharmacokinetic study of Uncaria alkaloids. The pharmacokinetic results showed that the six Uncaria alkaloids metabolized rapidly in mice with a half-life between 0.6 and 4.4 h. The bioavailability study also showed satisfactory oral absorption of each alkaloid [32]. Similar studies were performed on other different herbal medicines [33–35]. **Table 3** depicts typical examples of application of UHPLC in pharmacokinetics and bioavailability studies whereby chromatographic and mass settings are crucial for the method's high sensitivity and short retention period in addition to the extraction and sample preparation stages. With modern technology, such as ultra-performance liquid chromatography based on tandem mass spectrometry, it is now possible to detect drugs with great sensitivity and speed, especially for counterfeit medications with variations in the package or appearance. UHPLC–MS/MS analytical method development was used to investigate the rate of degradation of pharmaceuticals in microbial fuel cells (MFC). Additionally, by observing the voltage produced by the microorganisms and the rate of chemical oxygen demand (COD) elimination, the capability of MFC to treat urine spiked with medications was examined. The performance of MFC and medicinal additives was shown to be related, according to the findings [36].

A study done on Pimavanserin indicates that it passes the blood–brain barrier and approaches a Cmax of 21.9±6.66 ng/g in 2.0 hours, according to pharmacokinetics and brain uptake experiments using the technique. Additionally, it was discovered that the ratio of brain to plasma for pimavanserin (Kbrain/plasma) is 0.16±0.05 and that it is quickly removed. The developed method using UHPLC was linear ($R^2 > 0.99$ over the range of 0.1–300 ng/mL in plasma and 0.25–300 ng/g) in the brain homogenate [39].

Experimental Conditions	Application	Analytes	Sample matrix	Type of column	Oven T°c	Flow rate	MP	Finding	Ref.
UHPLC–MS-MS	Determination of Diphenidol (DPN)	DPN	Human Plasma and pork liver	β	at 4°C	0.3 mL/min	β	Detection and quantification of DPN in 15 real forensic cases	[27]
UHPLC–MS/MS	Assay for the quantification of tofacitinib (TOF)	TOF	Human plasma	∞	40°C	0.25 mL/ min	∞	TOF in human plasma samples was determined	[28]
UHPLC–MS/MS	Simultaneous quantitation of the drug and metabolites	Vorolanib	Human plasma	Ω	40°C	0.40 mL/min	Ω	UHPLC–MS/MS method was developed and validated	[29]
UHPLC–MS/MS	TDM	9 antiepileptic drugs	Human plasma	Λ	40°C	0.40 mL/min	^	3 antiepileptics were not within the effective therapeutic range	[30]

UHPLC Column: β ACQUITY UHPLC BEH C18 column (100 mm × 2.1 mm, 1.7 μm). ∞UHPLC BEH C18 (50 × 2.1 mm, 1.7 μm). Ω Acquity BEH C18 column. Λ Agilent Eclipse XDB-C18 column (50.0 × 2.1 mm, 1.7 μm), Mobile Phase: β (Mp-A)20 mmol/L ammonium acetate, 0.1% formic acid in water and 5% acetonitrile. (Mp-B) was acetonitrile. € acetonitrile: 40% tetrabutylammonium hydroxide., ∞ acetonitrile and 10.0mM ammonium acetate. Ω (MP-A) 0.1 % formic acid in deionized water, (MP-B) was composed of 0.1 % formic acid in HPLC-grade acetonitrile. Λ water and acetonitrile. @ Methanol: water (95:5, v/v).

Table 2.
Application of UHPLC in analysis of drugs in human plasma.

Table 3 describes the utilization of UHPLC for the study of pharmacokinetics of wide array of human medicines where the method was accurate, sensitive, and with a short analysis time.

4.4 Identification of metabolites (metabolomics)

Metabolomics a field of studies dealing with identification of many different metabolites that give virtual devices a far-reaching opportunity of utilizations in the fields of pharmacology, toxicology, enzyme discovery, and systems biology [43]. It intends to identify and quantify the full complement of low-molecular-weight, soluble metabolites in actively metabolizing tissues [44].

It also has an extensive contribution in improving ones understanding of disease mechanisms and drug effects. It improves our ability to predict individual variation in drug response phenotypes. Substantial attention has been developed in the application of metabolomics to describe different pathological states of human diseases such as cancer, diabetes, autoimmune, and coronary diseases, among others [43]. Recent technological advances allow characterizing plenty of metabolites from a small quantity of a biological samples. Numerous experiments conducted on cells and tissue cultures played a significant role in improving our biomedical understanding. This at the same time provides a foundation to interpret the concerns related to metabolic processes of the test subject [45]. UHPLC is considered suitable for metabolite profiling and metabolomics among the various LC platforms, especially for large-scale untargeted metabolic profiling due to its sensitivity, selectivity, and enhanced reproducibility [44]. The sensitivity and selectivity of UHPLC at low detection levels produces a precise and reliable data that can be used for a variety of reasons including pharmacokinetics, toxicity, and bioequivalence studies of different metabolites [46].

A metabolism study is considered to be mandatory in a drug development process. It has a vital role in the development of a new chemical entity (NCE). It aids in identifying the active metabolites in so doing, monitoring the possibility of reactive metabolite formations, augmenting pharmacokinetic and pharmacodynamics properties, comparing preclinical with clinical metabolism profile, understanding clearance, and predicting drug–drug or food-drug interactions at early stages of drug discovery [47]. By the time a NCE reaches the development stage, its metabolites' characterization becomes a critical process. The feeble spots of metabolites of drug candidate molecules are recognized and protected by changing the compound structure early [5]. Various analytical techniques have been coupled with UHPLC for the study of metabolomics. From the possible coupled techniques with UHPLC, nuclear magnetic resonance, the most uniform detection technique, has a lower sensitivity compared to mass spectrometry (MS)—and thus, the detection ability of low-abundance metabolites is restricted. Therefore, high-resolution mass spectrometry (HR-MS) and tandem mass spectrometry (MS/MS) are widely used in metabolomics. HR-MS, such as quadrupole time-of-flight mass spectrometry (QTOFMS), provides accurate mass and specific fragment patterns of MS/MS, which can improve the speed and the efficiency of metabolite identification [48]. Higher sensitivity, greater resolution, and faster separation are all benefits of using UHPLC coupled with quadrupole time of flight tandem mass spectrometry (UHPLC-QTOF/MS) for the evaluation of modified metabolites in intricate components. With the assistance of UHPLC-QTOF-MS/MS, a total of eight metabolites of a newly discovered piperazine-based anticancer molecule (IMID-2) were found and described in various matrices

Experimental Condition	Application	Analytes	Sample matrix	Types of column	Column T°c	Flow rate	MP	Finding	Ref.
UHPLC-MS/ MS	Removal efficiency of pharmaceuticals	Trimethoprime	Urine	*	30°c	0.3 mL/min	*	Removal rate achieved was 96±2 %	[36]
UHPLC-MS/ MS	Pharmacokinetic study	Celecoxib, dezocine and dexmedetomidine	Beagle plasma	@	45°c	0.4 mL/min	@	UHPLC-MS/MS method was established.	[37]
UHPLC-MS/ MS	Pharmacokinetics	Narciclasine, 7-deoxynarciclasine	Rat plasma	Λ	-	0.4 mL/ min	^	Pk of narciclasine and 7-deoxynarciclasine.	[38]
UHPLC-MS/ MS	Pharmacokinetics and brain uptake studies in mice	Pimavanserin	Mice brain and plasma	&	-	0.25 ml/min	&	Pimavanserin penetrates the blood-brain barrier	[39]
UHPLC-MS/ MS	Pharmacokinetics	Amiodarone, desethylamiodarone, dronedarone, desbutyldronedarone	Rat plasma	€	40°c	0.30 mL/min	€	UHPLC-MS/MS method was developed and PK was described	[40]
UHPLC–MS/ MS	Pharmacokinetics Study of Eupatilin	Eupatilin (flavone derived from Artemisia plants)	Rat Plasma	£	40°c	0.4mL/min	£	Pk of eupatilin was determined in rats.	[41]
UHPLC–MS/ MS	Determination of ledipasvir, sofosbuvir and its metabolite	Ledipasvir, sofobusvir,	Plasma	¥		0.40 mL/min	¥	Pharmacokinetic investigations of ledipasvir, sofosbuvir in rats	[42]

*UHPLC column: *The Agilent C-18 (2.1 294 mm × 100 mm, 1.8 μm) Zorbex Eclipse plus column for Levofloxacin, @Acquity UHPLC BEH C18 column (2.1 mm × 50 mm, 1.7 μm). Λ A UHPLC HSS T3 (2.1 mm × 100 mm, 1.8 μm), &UHPLC BEH™ C18 column (50 mm × 2.1 mm i.d., 1.7 μm. €Acquity UHPLC BEH C18 (2.1 mm x 50 mm, 1.7 μm). £ BEH C18 column (2.1 mm × 100 mm, 1.7 μm). ¥ Acquity BEH C18 column (2.1 mm × 50 mm, 1.7 μm). Mobile Phase: *0.5% FA in 1mM ammonium acetate and MeOH (1:1).@Acetonitrile (Mp-A) −0.1% formic acid aqueous solution (Mp-B). # Methanol (Mp-A) and 5 mM ammonium formate in water (Mp-B). $ water with 0.1% formic acid (Mp-A) and acetonitrile (Mp-B). Λ acetonitrile and water (with 0.1% formic acid). & (Mp-A) 0.1% formic acid in acetonitrile and (Mp-B) 0.1% formic acid in 20 mM ammonium acetate. € (Mp-A) acetonitrile) and (Mp-B) 0.1% formic acid aqueous solution. £ (Mp-A) acetonitrile and (Mp-B) water with 0.1% formic acid). ¥ Mp-A) acetonitrile), and (Mp- B) 0.1% formic acid.*

Table 3.
Application of UHPLC in pharmacokinetics and bioavailability studies.

including in rat liver microsomes (RLM), human liver microsomes (HLM) and rat S9 fraction (RS9), rat plasma, urine, and feces [47, 49]. Untargeted metabolomics in diseased plants may offer a fresh viewpoint and advance our comprehension of plant defense mechanisms. A relevant method for comparing plant metabolite changes is metabolomics. UHPLC paired with HR-MS is the most extensively used metabolomics technology due to its great sensitivity [48].

Table 4 depicts examples in the study of metabolomics using UHPLC. One study investigated and identified the metabolic characteristics of serum samples from children with urolithiasis and normal controls through UHPLC–MS-based metabolomics study approach. Forty differential metabolites were identified, mainly involved in retinol metabolism, steroid hormone biosynthesis, and porphyrin and chlorophyll metabolism. These results indicated that the metabolic phenotype of serum in patients with kidney stones was significantly different from that found in normal controls. The study provided a new understanding into the potential pathogenesis of urolithiasis, which may help to develop novel therapeutic strategies and preventive interventions [53]. Metabolomics can also be used to identify and quantify metabolites from plants. UHPLC–MS method was developed for chemotaxonomic study of seed accessions belonging to 16 different species of ***Vicia*** (vetch species family Fabaceae-Faboideae). These seeds are produced and consumed worldwide for their nutritional value. Both domesticated and wild taxa were analyzed by chemometrics-based UHPLC–MS method. A total of 89 metabolites were observed in the examined ***Vicia*** accessions. Seventy-eight out of the 89 detected metabolites were annotated. The study shows UHPLC–MS metabolomics method ability to discern the diversity of metabolites at the intrageneric level among ***Vicia*** species [54].

The discoveries of metabolic pathways and metabolites could provide a certain theoretical basis for drug discovery and pharmacological mechanism illucidation research [50].

4.5 Detection of impurities

Identification of impurities in raw materials and the final products is one of the most vital stages of a drug development process [4, 55]. Regulatory authorities give significant attention to impurity profiling. Starting materials, intermediates, precursors, etc., are the most common impurities found in every API unless proper care is taken in every step involved throughout the multi-step synthesis. Sometimes, impurities of intermediates and precursors generate structurally related by-products during synthesis [56].

HPLC with sufficient resolution has been providing an excellent detection and determination of the lowest level of impurities with highly reproducible results. However, due to the presence of excipients, there is prolonged HPLC analysis time so it becomes necessary to perform several analytical runs to get the required data time. Hence, the UHPLC/MS technique is operational at alternate low and high collision energies. The fast change of collision energy produces both precursor and product ions of all analytes present in the sample, which allows rapid identification and profiling of impurities [5]. **Table 5** shows several methods for determining, quantification, and characterization of pharmaceutical, impurities using UHPLC have been devised that are easy to use, quick, appropriate, precise, and accurate. In terms of specificity, system suitability, linearity, limit of detection and quantitation, accuracy, precision, robustness, and solution stability, the suggested technique was fully

Experiment Condition	Application	Analytes	Matrix	Type of column	Oven T°C	Flow rate	MP	Findings	Ref
UHPLC-QTOF-MS/MS	Metabolite profiling	2-[4-(3,4-dichlorophenyl) piperazin1-yl]-1-(imidazo[1,2-a]pyrimidin-3-yl)ethanone} (IMID-2)	Blood, urine and feces of rats	Λ	Ambient	0.4 ml/min	Λ	8 metabolites of IMID-2 were identified	[47]
UHPLC-QTOF-MS	Identification of Metabolites	Rice Leaves	Powdered leaves	#	25°C	0.4 mL/min	#	Secondary metabolites are prostanoids	[48]
UHPLC/Q TOF MS	Metabolic Study of Stable Isotope	Indolinone Derivative compounds	Hepatic cell line	&	45°C	-	&	Seven metabolites were discovered	[50]
UHPLC-QTOF/MS	Metabolomic study from subjects after long-term occupational exposure to low concentration acrylamide	Acrylamide (ACR)	Urine and serum	*	35°C	-	*	New biomarkers of nervous system injury	[49]
UHPLC-Q-TOF/MS	Hepatotoxicity of Emodin and Detoxification of Dihydromyricetin (DMY)	Ampelopsis grossedentata (TCM)	Serum, urine, and liver samples from rats	@	40°C	0.4 mL/min	@	Liver protective effect of DMY	[51]
UHPLC-Q-TOF/MS	To analyze the metabolites in the urine of cardiovascular Heart disease patients	Metabolities associated with CHD	Urine	$	45°C	0.3 ml/min	$	14 biomarkers associated with CHD	[52]

*UHPLC column: Λ SB C_{18} column of (100 x 3mm, 1.8μm). # YMC-Triart C18 column (2.0 × 150 mm, 1.9 μm). & HSS T3 column (100 × 2.1 mm, i.d., 1.8 μm). *Acquity UHPLC™ BEH C18 column (100 mm × 2.1 mm i. d., 1.7 μm). @Acquity UHPLC HSS C18 column (1.7 μm, 2.1 × 100 mm). $ Acquity UHPLC BEH C18 column (2.1 mm × 100 mm, 1.7 μm, Waters). Mobile Phase: Λ Ammonium acetate (Mp-A) and acetonitrile (Mp-B). # 0.1% formic acid in water (Mp-A) and 0.1% formic acid in ACN (Mp-B). & (Mp-A) 0.02% formic acid in ammonium acetate (5 mM) and acetonitrile (Mp-B). * (Mp-A) was 0.1% formic acid in water and (Mp-B) was 100% acetonitrile.@0.1% formic acid–water (Mp-A), (Mp-B) 0.1% formic acid acetonitrile. $ 0.1% formic acid in water (Mp-A) and 0.1% formic acid in acetonitrile in (Mp-B).*

Table 4.
Use of UHPLC in identification of metabolites.

Experimental Conditions	Application	Analytes	Sample matrix	Type of column	Column T⁰	Flow rate	Mp	Finding	Ref.
UHPLC	Determination of process related impurities	Azathioprine and related impurities	Solution of drug and impurities	Λ	30°C	0.35 mL/ min	Λ	All the impurities were well resolved within 5 min	[56]
UHPLC-QTrap-MS/MS	Characterization of drug-related impurities	Daptomycin	API solution	*	15°C and 55°C	0.3 mL/min	*	Characterization of drug-related impurities	[57]
UHPLC with QToF-MS/MS	quantification of rosuvastatin (RSV) and its related impurities	RSV and related impurities	RSV dosage form	@	40°C	0.3 mL/min	@	The developed method was validated	[58]
UHPLC-QDa	Detection of potential impurities	Daclatasvir	Drug solution	#	35°C	0.4 mL/ min	#	unknown degradation products were identified	[59]
UHPLC	Determination of impurities in pharmaceutical dosage forms	Glicazide	Forced degradation products	$	45°C	0.7 mL/min	$	UHPLC method was developed and validated	[60]
RP-UHPLC	To determine the drugs and impurities	Mebendazole, quinfamide, its impurities	Drug sample solution	α	40°C	0.38 mL/min and 2μL	α	The method was validated	[61]
UHPLC	Quantitative determination of impurities in drug	(Enzalutamide) ENZ	ENZ soft gel capsule made into solution	π	40°C	0.2 mL min−1	π	Method based developed	[62]

*UHPLC Columns: Λ Acuity UHPLC BEH C18 (100 2.1 mm, 1.7 μm). * BEH C18 Van Guard precolumn (2.1 5 mm, 1.7 μm) and CSH C18 column (150 2.1 mm, 1.7 μm). @BEH C-18 100 x 2.1 mm, 1.7 μm. # BEH phenyl 100 × 2.1 mm, 1.7-μm. $CSH-C18 column (50 mm × 2.1 mm × 1.7 μm), Acquity BEH-C18 column (50 mm × 2.1 mm × 1.7 μm), Phenomenex Kinetex-C18 column (50 mm × 2.1 mm × 1.7 μ). α Cortecs UHPLC C18, 150 mm x 2.1 mm, 1.6 μm. π ACQUITY CSH C18 (100 × 2.1 mm × 1.7 μm). Mobile Phase: Λ (Mp-A) 0.05% trifluoroacetic acid in water and acetonitrile (Mp-B).*(Mp-A) was 0.2% formic acid and 0.03% ammonia in water, and (Mp-B) was 0.2% formic acid in acetonitrile.@ MeOH and 0.1% TFA in the ratio of 50:50.#(Mp-A) with sodium perchlorate monohydrate and 1-octanesulfonic acid sodium salt with perchloric acid solution and (Mp-B) sodium perchlorate and 1-octanesulfonic acid sodium salt..$ 5mM ammonium acetate buffer of pH 4 and 10% ammonium acetate buffer + 90% acetonitrile in the ratio of 65:35. α (Mp-A) 0.025% Trifluoracetic acid and (Mp-B) BAcetonitrile: Methanol (90:10v/v). π potassium phosphate monobasic buffer and acetonitrile (10 mM, adjusted to pH 4.0 with 1% orthophosphoric acid.*

Table 5.
Application of UHPLC in detection of impurities.

validated. The validation results also indicated positive data for each of the examined parameters [56–63].

4.6 Rapid analysis of dosage formulations

Since disease-causing microorganisms are in more mutation than ever and humans' lifestyle is causing more illness than before, the pharmaceutical industry is under intense pressure to increase productivity and bring new drugs onto the market in a short period. UHPLC system provides accurate and reproducible results in rapid isocratic and gradient methods for drug molecules analysis in dosage forms [34]. This in turn helps in improving pharmaceutical manufacturing efficiency. **Table 6** shows examples of UHPLC application in the analysis of dose formulation.

4.7 Food safety

Food products' safety and quality are a concern for consumers and governments. Analytical information, including surveillance data for both recognized and newly identified contaminants, is also indispensable. However, information about food contamination incidence is still limited [65].

Food additives and mycotoxins are among the major toxic treats in different foods consumed worldwide [64]. Food stuffs are vulnerable to infection and contamination in the field or during storage. Mycotoxins, for instance aflatoxins (Afs), Afs B1 is the most common aflatoxin, and is the most toxic and carcinogenic and ochratoxin A (OTA) along with OTA C is known to cause hepatotoxicity, immunotoxicity, neurotoxicity, teratogenicity, and carcinogenicity. The challenge imposed on food safety due to mycotoxins is given currently a rising attention by the government, regulatory authority, and academia globally [66].

The safety of food components and contaminants can only be guaranteed when a good analysis approach is available. During the past few decades, chromatography has been recognized as one of the methods employed to identify and quantify food contaminants. For both qualitative and quantitative research, this novel procedure enables the isolation, purification, and detection of components from a mixture. UHPLC–MS has lately been utilized to estimate food pollutants and components in order to improve food safety [64].

Sensitivity of UHPLC has reached ppb and ppt levels; thus by these quality analytical results, a food analyst would be more confident in ensuring safe food consumption [65]. It is also ideal for analyzing low-level concentrations of food additives where high sample throughput is required without affecting the method's accuracy and sensitivity [67]. **Table 7** shows applying the rules in the context of complicated matrices, like coffee, which frequently necessitates costly and time-consuming strategies. Without clean-up, the UHPLC–MS/MS approach offered adequate sensitivity and resolution. The performance attributes of the approach include high LOQ, recovery, and precision [66]. Additionally, the UHPLC-ESI-MS/MS technology is utilized to discover other potential microbial metabolites present in samples and to confirm the identity of mycotoxins that have been detected [68]. UHPLC has also been employed to detect adulterated milk powder samples, with an acceptable linear correlation coefficient. It is used to assess food products and their diversity or assure their quality and authenticity [69, 70]. Quantification of essential nutrients from human milk was quantified by UHPLC to direct the diet for lactating women [72].

DOI: http://dx.doi.org/10.5772/intechopen.110540

Experimental Condition	Application	Analytes	Sample matrix	Type of column	Oven To	Flow rate	MP	Finding	Ref.
UHPLC-MS	Rapid analysis of dose formulation	Mefenamic acid (MFA) and chloramphenicol (CA)	Aliquots (1 mL) of the dose formulation	Λ	55 °C	0.1 to 1.2 mL/min	Λ	Analysis of CA and MFA in the dose formulation was done	[64]

*UHPLC Columns: Λ BEH C18 column (2.1 x 50 mm, 1.7 m). Mobile Phase: Λ Water/acetonitrile/formic acid (65/35/0.1; v/v/v). * a buffer with pH 4.5 and acetonitrile (60:40% v/v).*

Table 6.
Applications of UHPLC in analysis of dosage formulations.

Experimental condition	Application	Analytes	Sample matrix	Type of column	Oven T°c	Flow rate	MP	Findings	Ref.
UHPLC-MS/MS	Analysis of aflatoxins and ochratoxin A (OTA)	Raw coffee sample	Aflatoxin B1 aflatoxin B2 (aflatoxin G1 aflatoxin G2 and OTA	β	45°C	0.45 mL/min	β	OTA was found in 3 out of 4 samples commercialised in Morocco.	[66]
UHPLC –MS/MS and UHPLC–QTOF-MS	Determination of multiple mycotoxins	Mycotoxins	Feed stuffs	@	25°C	0.35 mL/min	@	Non-target mycotoxins/ fungal metabolites have been identified	[68]
UHPLC-PDA	Anthocyanins profiling in *Vaccinium* L. berries	Berries	Extracts of bilberries, cranberries and lingon berries	#	30°C	0.5 mL/min	#	Separation and detection of major and minor 406 anthocyanins	[69]
UHPLC	Detection of adulterated camel milk powder	Commercial camel milk powder product	Whey protein solution	α	30°C	0.2 mL/ min	α	8 samples were found to be adulterated with a high level of cow milk	[70]
UHPLC–MS	Analysis of water-soluble azo dyes in soft drinks	Water soluble azo dyes	Soft drinks	Λ	35°C	0.3 ml min1	Λ	Simultaneous determination of 11 water-soluble azo dyes in soft drinks	[71]
UHPLC-MS/MS	Analyzing B-vitamins in Human Milk	Vitamin B's	Breast Milk	Ω	40°C	0.35 ml/min	Ω	Fast quantification of vitamin B's has been developed and validated	[72]
UHPLC-QTOF-MS	Quantitative determination of 20 antimicrobial residues	Tetracyclines, quinolones, sulfonamides and diaminopyrimidines	Edible muscle plus skin tissue of European sea bass	∞	30°C	0.4 mL/min	∞	The proposed method was applied for the analysis of contaminated fish samples	[73]

UHPLC Columns: β ACQUITY UHPLC® BEH C18 analytical column (1.7 μm, 2.1 mm × 100 mm). @ACQUITY UHPLC BEH C18 column (50 × 2.1 mm, 1.7 μm particle size). # ACT, Aberdeen, UK; 100×2.1 mm, 1.7 μm particle size. α BEH 300 C4 column (100 mm x 2.1 mm i.d., 1.7 μm). Λ Acquity UHPLC™ BEH Shield RP18 (2.1 100 mm, 1.7 μm. Ω HSS T3 (2.1 x 100 mm, 1.8 μm). ∞ a C18 BEH column (50 mm × 2.1 mm, 1.7 μm, Waters). Mobile Phase: β (Mp-A)(H2O with 10 mM HCOONH4, pH = 5) and eluent B (MeOH with 10 mM HCOONH4, pH = 5). @(Mp-A)water containing ammonium formate (0.15 mmol/L) and formic acid (0.1%) and MeOH(Mp- B). # (Mp-A) aqueous 10% formic acid solution and (Mp-B) acetonitrile.α(Mp-A)ultra-pure water containing 0.1% (v/v) trifluoroacetic acid (TFA); eluent (Mp-B) was acetonitrile containing 0.09% (v/v) TFA. ΛMethanol and (Mp-B)0.1% (v/v) NH3 and H2O in water (Mp-B). Ω (Mp-A) Acetontrileand (Mp-B) 2.5mmol/L ammonium formate aqueous solution. ∞(Mp-A)0.1% aqueous FA and (Mp-B) methanol with 0.1% formic acid.

Table 7.
Application of UHPLC in food safety determination.

4.8 Application in agricultural sector

UHPLC has also been useful in the agricultural sector for the study of soil components, pesticide residue analysis, and crop analysis. Agricultural products contain not only plant materials but animal products too. Thus, safety and quality of such products need to be studied and quality ensured before consumption [74, 75]. In this direction, a rapid, sensitive, and specific ultra-high-performance liquid chromatography tandem mass spectrometry (UHPLC–MS) method was developed for the analysis of tetracycline antibiotics, including tetracycline (TC), oxytetracycline (OTC), chlortetracycline (CTC), and their 4-epimers (4-epi TCs) in agricultural soil fertilized with swine manure. The limits of detection for the soil extraction method ranged from 0.6 to 2.5 μg kg^{-1} with recoveries of 23.3–159.2%. The method was applied on an agricultural field in an area with an intensive pig-fattening farm. Tetracyclines detected in the soil varied from 2.8 to 42.4 μg kg^{-1}. The results made evident that soil from swine farms can be rigorously contaminated with tetracycline antibiotics and their metabolites [74]. **Table 8** summarizes the effect of fungicide and pesticide treatment on the agricultural lands in which case they were considered and quantified by analyzing the harvested products using UHPLC. Besides seasonal variations on the harvested food, both occurrence and concentration of antioxidant abilities and flavonoids were investigated using UHPLC [75–77].

4.9 Method development and validation

UHPLC has been used broadly for method development and validation purpose. UHPLC plays a crucial role in fundamental laboratory function by increasing efficiency, reducing costs, and improving opportunities for business success. Using UHPLC, analysis times become as short as one minute, methods can be optimized in just one or two hours, thereby appreciably decreasing the time required to develop and validate new analytical method [79]. As an example, a study done for the simultaneous estimation of paracetamol and caffeine capsules dosage form indicated that a method was validated using various validation parameters such as accuracy, precision, linearity, and specificity. Moreover, the results show that the method could find practical application as a quality control tool for analyzing the drug in its capsule dosage forms in pharmaceutical industries offering the above-mentioned advantages [80].

5. Advancements of UHPLC

UHPLC is thought to provide higher rates of efficiency, sensitivity, speed, and resolution of UHPLC devices make the system perfect for use with mass spectrometer. More laboratories are finding UHPLC–MS systems to be practicable due to the accessibility and low cost of a new generation of MS equipment [81]. The software used and created for UHPLC systems, which makes the instrument easy to manage, diagnose, and monitor, is primarily responsible for another breakthrough. This makes a significant contribution to the dynamic data processing and information management that convert the findings of the UHPLC system into useful knowledge. In addition, the recording technology found on the majority of UHPLC columns allows for the recording of column history. Another important development is the availability of sample organizers that multiply system capacity by more than 10 times. Additionally,

Experimental condition	Application	Analytes	Sample matrix	Type of column	Column T°	Flow rate	MP	Finding	Ref.
UHPLC-IMS-QToF	Impacts of Fungicide Treatment	Potato samples	Potato metabolites and extracted pesticides	*	55°C	0.45 mL/min	*	Fungicide usage on the potato tuber metabolome is higher	[75]
UHPLC	Seasonal variations in phenolic compounds	*Cornus stolonifera*	Plant material	@	42°C	0.6 mL /min	@	High phenolic content with profound antioxidant activity	[76]
UHPLC/MS-MS	Analytical detection technology	Pesticide residue	fruits, vegetables and grains	#	350°C	0.24mL/min	#	8 pesticide residues were determined	[77]
UHPLC-QTof-MSE M	Metabolite changes in under drought stress.	The hop (*Humulus lupulus L.*)	leaf samples	Λ	40°C	0.5 mL/min	Λ	Drought treatments produced qualitatively distinct changes	[78]

*UHPLC Coulmns: * UHPLC BEH C18 column (150 × 2.1 mm , 1.7 μm).@ 2.1*100 mm, 1.7 μm ACQUITY UHPLC BEH C18. # UHPLC BEHC18, 1.8 μm, 2.4 mm×55 mm. Λ UHPLC BEH C18, 1.7 μm, 2.1 × 100 mm column attached to an ACQUITY UHPLC BEH C18, 1.7 μm, 2.1 × 5 mm VanGuard precolumn. Mobile Phase: *(Mp-A) Water and (Mp-B) acetonitrile. @(Mp-A) 0.3% phosphoric acid, 5% methanol in 18.2 MΩ/cm water; and (Mp-B) acetonitrile. # (Mp-A) 4.5mmol/L ammonium acetate solution containing 0.2% formic acid (Mp-B) is acetonitrile. Λ (Mp-A) water and (Mp-B),acetonitrile.*

Table 8.
Application of UHPLC in agricultural sector.

some businesses offer a virtual technical support service, which enables them to give customers prompt, proactive help that meets their needs to the fullest extent [14].

Examples of organizers that boost system capacity more than ten times are technologically availed. Thus, UHPLC–MS systems are becoming more practical for more laboratories [82, 83]. Despite its expensive price, these developments will further elevate the UHPLC as a crucial analytical tool for sample studies that will be the analysts' first choice worldwide. Future research on engineering and material science topics might also be able to resolve this issue, making this analytical instrument more accessible to many laboratories.

6. Conclusion

Liquid chromatography has been replaced by UHPLC in a novel approach of improvement. It is a sort of separation technique with nearly identical principles as HPLC. The UHPLC uses small packing particles (less than 2 μm in size), which directly influence the length of the column and, in turn, minimize solvent consumption and shorten analysis times. The quick analysis makes it possible to do several analytical tasks quickly. Additionally, UHPLC has been a popular option for analytical work in recent years due to its better sensitivity and resolution. It is used in many fields because of the excellent quality outcomes that are produced. In addition, UHPLC had a wide range of applications in the fields of bioavailability, pharmacokinetics, natural medicines, metabolomics, food safety, and agricultural sectors. The pharmaceutical industry is benefiting most in the drug discovery process. With the aid of UHPLC, researchers may now complete a range of investigations in a short duration of time while saving money on analysis.

Author details

Feruza Ahmed[1], Tadele Eticha[1], Ariaya Hymete[1] and Ayenew Ashenef[1,2]*

1 Department of Pharmaceutical Chemistry, School of Pharmacy, College of Health Sciences, Addis Ababa University, Addis Ababa, Ethiopia

2 Center for Innovative Drug Development and Therapeutic Trials for Africa (CDT-Africa), College of Health Sciences, Addis Ababa University, Addis Ababa, Ethiopia

*Address all correspondence to: ayenew.ashenef@aau.edu.et

References

[1] Chopade SV, Patil VR. Introduction to new chromatography technique - UPLC. Available from: https://www.pharmatutor.org/articles/new-chromatographic-technique-uplc-ultra-performance-liquid-chromatography [Accessed: December 13, 2021].

[2] Gumustas M, Kurbanoglu S, Uslu B, Ozkan SA. UPLC versus HPLC on drug analysis: Advantageous, applications and their validation parameters. Chromatographia. 2013;**76**:1365-1427. DOI: 10.1007/s10337-013-2477-8

[3] Chawla G, Ranjan C. Principle, instrumentation, and applications of UPLC: A novel technique of liquid chromatography. Open Chemistry Journal. 2016;**3**(1):1-16. DOI: 10.2174/1874842201603010001

[4] Klimczak I, Gliszczynska wiglo A. Comparison of UPLC and HPLC methods for determination of vitamin C. Food Chemistry. 2015;**175**:100-105. DOI: 10.1016/j.foodchem.2014.11.104

[5] Kumar A, Saini G, Nair A, Sharma R. Review uplc: a preeminent technique in pharmaceutical analysis. Acta Poloniae Pharmaceutica. 2012;**69**(3):371-380

[6] Chesnut SM, Salisbury JJ. The role of UHPLC in pharmaceutical development. Journal of Separation Science. 2007;**30**(8):1183-1190. DOI: 10.1002/jssc.200600505

[7] Singh S. A brief review on ultra performance liquid chromatography. World Journal of Pharmacy Science. 2015;**6**(15):407-422. DOI: 10.20959/wjpr201715-10136

[8] Taleuzzaman M, Ali S, Gilani SJ, Imam SS, Hafeez A. Ultra performance liquid chromatography (UPLC) - A review. Austin Journal of Analytical and Pharmaceutical Chemistry. 2015;**2**(6):1056

[9] Swartz ME. UPLC: An introduction and review. Journal of Liquid Chromatography & Related Technologies. 2005;**28**:37-41. DOI: 10.1081/JLC-200053046

[10] Goswami S. UPLC TM-A review on its recent advances in instrumental analysis UPLC TM-A review on its recent advances in instrumental analysis. Indian Journal of Natural Sciences. 2021; **12**(68):1-18

[11] Bhanot D. What are the Benefits of UHPLC over normal Analytical HPLC? [Internet]. 2014. Available from: https://lab-training.com/2014/04/17/what-are-the-benefits-of-uhplc-over-normal-analytical-hplc/ [Accessed: December 2021].

[12] Chandraman K. An updated review on ultra performance liquid chromatography. Analytical Chemistry: An Indian Journal. 2016;**16**(15):114

[13] Shah K. Ultra performance liquid chromatography (Uplc): a modern chromatography techique. Pharma Science Monitor. 2014;**4**(3):23

[14] Naresh K, Bhawani S, Kumar TM. Ultra performance liquid chromatography. International Journal of Pharma Medicine and Biological Sciences. 2014;**3**(3):84-94

[15] Agilent Technologies. Agilent 1290 Infinity Evaporative Light Scattering Detector (ELSD) For Low Molecular Weight Compounds. 2012; 8. Available from: https://www.agilent.com/cs/library/datasheets/public/5990-8180EN.pdf

[16] Setyaningsh W, Palma M, Barroso CG. Comparison of HPLC and UPLC

methods for the determination of melatonin in rice. In: Nesměrák K, editor. Proceedings of the 11th Modern Analytical Chemistry, Charles University in Prague, Faculty of Science, Czech Republic. 2015. pp. 170-175. DOI: 10.13140/RG.2.1.2300.1687

[17] Morais DR, Rotta EM, Sargi SC, Schmidt EM, Bonafe EG, Eberlin MN, Sawaya A.C.C.H.F and Visentainer J. Antioxidant activity, phenolics and UPLC-ESI(−)-MS of extracts from different tropical fruits parts and processed peels. Food Research International 2015;77:392–399.DOI: 10.1016/j.foodres.2015.08.036

[18] Jeong SH, Jang JH, Lee GY, Yang SJ, Cho HY, Lee YB. Simultaneous determination of fourteen components of Gumiganghwal-tang tablet in human plasma by UPLC-ESI-MS/MS and its application to pharmacokinetic study. Journal of Pharmaceutical Analysis. 2021;**11**(4):444-457. DOI: 10.1016/j.jpha.2020.08.003

[19] Wang N, Yang B, Zhang J, Zheng Y, Wang S, Zhang X, et al. Metabolite profiling of traditional Chinese medicine XIAOPI formula: An integrated strategy based on UPLC-Q-Orbitrap MS combined with network pharmacology analysis. Biomedicine & Pharmacotherapy. 2019;**2020**(121): 109569. DOI: 10.1016/j.biopha.2019.109569

[20] Li P, Liu S, Liu Q, Shen J, Yang R, Jiang B, et al. Screening of acetylcholinesterase inhibitors and characterizing of phytochemical constituents from Dichocarpum auriculatum (Franch.) W.T. Wang & P. K. Hsiao through UPLC-MS combined with an acetylcholinesterase inhibition assay in vitro. Journal of Ethnopharmacology. 2019;**245**:112185. DOI: 10.1016/j.jep.2019.112185

[21] Yuan L, Chen H, Ma X, Pan J, Gong ZP, Chen SY, et al. Herb-drug interaction: Application of a UPLC-MS/MS method to determine the effect of Polygonum capitatum extract on the tissue distribution and excretion of levofloxacin in rats. Hindawi. 2020; **2020**:2178656. DOI: 10.1155/2020/2178656

[22] Wang X, Zhang A, Zhou X, Liu Q, Nan Y, Guan Y, et al. An integrated chinmedomics strategy for discovery of effective constituents from traditional herbal medicine. Scientific Reports. 2015;**2016**(6):1-12. DOI: 10.1038/srep18997

[23] Wu P, Huang Z, Shan J, Luo Z, Zhang N, Yin S, et al. Interventional effects of the direct application of "sanse powder" on knee osteoarthritis in rats as determined from lipidomics via UPLC-Q-Exactive Orbitrap MS. Chinese Medicine. 2020;**15**(9):1-13. DOI: 10.1186/s13020-020-0290-5

[24] Chen M, Chen Y, Wang X, Zhou Y. Quantitative determination of talatisamine and its pharmacokinetics and bioavailability in mouse plasma by UPLC-MS/MS. Journal of Chromatography, B: Analytical Technologies in the Biomedical and Life Sciences. 2019;**1124**(March):180-187. DOI: 10.1016/j.jchromb.2019.06.015

[25] Van der Nagel BCH, Versmissen J, Bahmany S, van Gelder T, Koch BCP. High-throughput quantification of 8 antihypertensive drugs and active metabolites in human plasma using UPLC–MS/MS. Journal of Chromatography, B: Analytical Technologies in the Biomedical and Life Sciences. 2017;**1060**(June):367-373. DOI: 1016/j.jchromb.2017.06.029

[26] Gouget H, Noé G, Barrail-Tran A, Furlan V. UPLC–MS/MS method for the

simultaneous quantification of bictegravir and 13 others antiretroviral drugs plus cobicistat and ritonavir boosters in human plasma. Journal of Pharmaceutical and Biomedical Analysis. 2020;**181**:113057. DOI: 10.1016/j.jpba.2019.113057

[27] Yang L, Shen M, Xiang P, Shen B, Deng H, Qiang H, et al. Application of a validated UPLC–MS-MS method for the determination of diphenidol in biological samples in 15 authentic lethal cases. Journal of Analytical Toxicology. 2020; **45**:1-9. DOI: 10.1093/jat/bkaa160

[28] Bharwad KD, Shah PA, Shrivastav PS, Singhal P. Development and validation of a rapid and sensitive UPLC–MS/MS assay for the quantification of tofacitinib in human plasma. Biomedical Chromatography. 2019;**33**(4):1-9. DOI: 10.1002/bmc.4458

[29] Zheng X, Gao H, Zhang Y, Cui X, Jia R, Xue J, et al. Development of a rapid and sensitive UPLC–MS/MS assay for simultaneous quantitation of Vorolanib and its metabolite in human plasma and application to a pharmacokinetics study. Journal of Pharmaceutical and Biomedical Analysis. 2021;**199**:114034. DOI: 10.1016/j.jpba.2021.114034

[30] Qi Y, Liu G. A UPLC–MS/MS method for simultaneous determination of nine antiepileptic drugs in human plasma and its application in TDM. Biomedical Chromatography. 2021;**35** (7):e5090. DOI: 10.1002/bmc.5090

[31] Chen L, Weng Q, Li F, Liu J, Zhang X, YZ. Pharmacokinetics and bioavailability study of Tubeimoside I in ICR mice by UPLC-MS/MS. Acta Chromatographica. 2018;**30**(1):26-30. DOI: 10.1155/2018/9074893

[32] Chen L, Ma J, Wang X, Zhang M. Simultaneous determination of six Uncaria alkaloids in mouse blood by UPLC-MS/MS and its application in pharmacokinetics and bioavailability. BioMed Research International. 2020; **2020**:1030269

[33] Guo D, Dang J, Yang H, Fan Y, Cheng J, Shi Y, et al. Simultaneous determination of four flavonoids in rat plasma after oral administration of Malus hupehensis (Pamp.) Rehd. Extracts by UPLC-MS/MS and its application to a pharmacokinetics study. Journal of Pharmaceutical and Biomedical Analysis. 2020;**177**:112869. DOI: 10.1016/j.jpba.2019.112869

[34] Hu Y, Chen M, Wang Z, Lan Y, Tang L, Liu M, et al. Development of a validated UPLC–MS/MS method for determination of humantenmine in rat plasma and its application in pharmacokinetics and bioavailability studies. Biomedical Chromatography. 2017;**31**(12):e4017. DOI: 10.1002/bmc.4017

[35] Tong S, Zeng Y, Jianshe MA, Congcong WEN. Pharmacokinetics and bioavailability of liensinine in mouse blood by UPLC-MS/MS. Acta Chromatographica. 2021;**33**(4):333-337. DOI: 10.1556/1326.2020.00847

[36] Sharma P, Kumar D, Mutnuri S. Probing the degradation of pharmaceuticals in urine using MFC and studying their removal efficiency by UPLC-MS/MS. Journal of Pharmaceutical Analysis. 2021;**11**(3):320-329. DOI: 10.1016/j.jpha.2020.04.006

[37] Jie H, Xin-juan S, Si H-L, Song R-X, Zhang F, Qiu X-J, et al. Simultaneous determination of celecoxib, Dezocine and Dexmedetomidine in beagle plasma using UPLC-MS/MS method and the application in pharmacokinetics. Drug Design, Development and Therapy.

2021;**15**(6):2529-2541. DOI: 10.2147/DDDT.S314562

[38] Chen F, Yu Z, Wang X. Development of a UPLC-MS/MS method for the determination of narciclasine and 7-deoxynarciclasine in mouse blood and its application in pharmacokinetics. Journal of Chromatography, B: Analytical Technologies in the Biomedical and Life Sciences. 2021;**1180** (August):122899. DOI: 10.1016/j.jchromb.2021.122899

[39] Ezzeldin E, Iqbal M, Asiri YA, Ali AA, El-Nahhas T. A rapid, simple and highly sensitive UPLC-MS/MS method for quantitation of pimavanserin in plasma and tissues: Application to pharmacokinetics and brain uptake studies in mice. Journal of Chromatography, B: Analytical Technologies in the Biomedical and Life Sciences. 2020;**1143**(February):122015. DOI: 10.1016/j.jchromb.2020.122015

[40] Zhang Y, Zhu M, Xie S, Ye X, Xu X. Simultaneous determination of amiodarone, dronedarone, and their principal metabolites in SD rat plasma by UPLC-MS/MS and its application in pharmacokinetics. Arabian Journal of Chemistry. 2021;**14**(8):103300. DOI: 10.1016/j.arabjc.2021.103300

[41] Geng P, Luo X, Peng X, Lin Z, Chen W, Zhang J, et al. Development and validation of UPLC–MS/MS method for determination of eupatilin in rat plasma and its application in a pharmacokinetics study. Acta Chromatographica. 2018;**30** (4):231-235

[42] Pan C, Chen Y, Chen W, Zhou G, Jin L, Zheng Y, et al. Simultaneous determination of Ledipasvir, sofosbuvir and its metabolite in rat plasma by UPLC-MS/MS and its application to a pharmacokinetic study. Journal of Chromatography, B: Analytical Technologies in the Biomedical and Life Sciences. 2016;**1008**:255-259. DOI: 10.1016/j.jchromb.2015.11.056

[43] Nassar AF, Wu T, Nassar SF, Wisnewski AV. UPLC–MS for metabolomics: a giant step forward in support of pharmaceutical research. Drug Discovery Today. 2017;**22**(2):463-470. DOI: 10.1016/j.drudis.2016.11.020

[44] Xie G, Plumb R, Su M, Xu Z, Zhao A, Qiu M, et al. Ultra-performance LC/TOF MS analysis of medicinal Panax herbs for metabolomic research. Journal of Separation Science. 2008;**31**(6–7): 1015-1026. DOI: 10.1002/jssc.200700650

[45] Nawaz SS. The role of metabolomics in personalized medicine for diabetes. Report. 2021;**18**:501-508. DOI: 10.2217/pmc-2021-0083

[46] Rathod RH, Chaudhari SR, Patil AS, Shirkhedkar AA. Ultra-high performance liquid chromatography-MS/MS (UHPLC-MS/MS) in practice: Analysis of drugs and pharmaceutical formulations. Future Journal of Pharmaceutical Sciences. 2019;**5**(1):6. DOI: 10.1186/s43094-019-0007

[47] Panday NK, Thakkar D, Patel S, Shard A, Sengupta P. Metabolite profiling of IMID-2, a novel anticancer molecule of piperazine derivative: In silico prediction, in vitro and in vivo metabolite characterization using UPLC–QTOF–MS/MS. Biomedical Chromatography. 2021; **35**(6):1-11. DOI: 10.1002/bmc.5082

[48] Oh M, Park S, Kim H, Choi GJ, Kim SH. Application of uplc-qtof-ms based untargeted metabolomics in identification of metabolites induced in pathogen-infected rice. Plants. 2021;**10** (2):1-13. DOI: 10.3390/plants10020213

[49] Wang SY, Han D, Pan YL, Yu CP, Zhou XR, Xin R, et al. A urinary

metabolomic study from subjects after long-term occupational exposure to low concentration acrylamide using UPLC-QTOF/MS. Archives of Biochemistry and Biophysics. 2019;**2020**(681): 108279. DOI: 10.1016/j.abb.2020.108279

[50] Yang J, Zhang G, Wang Z, Meng J, Wen H. Metabolic study of stable isotope labeled Indolinone derivative in hepatocyte cell by UPLC/Q TOF MS. Journal of the American Society for Mass Spectrometry. 2021;**32**(6):1538-1544. DOI: 10.1021/jasms.1c00146

[51] Gao J, Shi N, Guo H, Gao J, Tang X, Yuan S, Qian J, and Wen B.UPLC-Q-TOF/MS-Based Metabolomics approach to reveal the hepatotoxicity of Emodin and detoxification of Dihydromyricetin, ACS Omega 2021;6(8):5348–5358. DOI: 10.1021/acsomega.0c05488.

[52] Huang M, Zhao H, Gao S, Liu Y, Liu Y, Zhang T, et al. Identification of coronary heart disease biomarkers with different severities of coronary stenosis in human urine using non-targeted metabolomics based on UPLC-Q-TOF/MS. Clinica Chimica Acta. 2019;**497**(July):95-103. DOI: 10.1016/j.cca.2019.07.017

[53] Wen J, Cao Y, Li Y, Zhu F, Yuan M, Xu J, et al. Metabolomics analysis of the serum from children with urolithiasis using UPLC-MS. Clinical and Translational Science. 2021;**14**(4):1327-1337. DOI: 10.1111/cts.12984

[54] Fayek NM, Mekky RH, Dias CN, Kropf M, Heiss AG, Wessjohann LA, et al. UPLC-MS metabolome-based seed classification of 16 Vicia species: A Prospect for Phyto-equivalency and chemotaxonomy of different accessions. Journal of Agricultural and Food Chemistry. 2021;**69**(17):5252-5266. DOI: 10.1021/acs.jafc.0c06054

[55] Kamal S, Sharad W. Step-up in liquid chromatography from HPLC to UPLC: A comparative and comprehensive review. The Pharma Innovation Journal. 2018;7(8):342-347

[56] Davadra PM, Mepal VV, Jain MR, Joshi CG, Bapodra AH. A validated UPLC method for the determination of process-related impurities in azathioprine bulk drug. Analytical Methods. 2011;**3**(1):198-204. DOI: 10.1039/c0ay00406e

[57] Liang X, Gou Z, Zhu X, Wang Y, Li N, Fu X, et al. Characterization of drug-related impurities using UPLC-QTrap-MS/MS in three scan modes: Application to daptomycin. International Journal of Mass Spectrometry. 2021;**468**:1-6. Article number: 116657. DOI: 10.1016/j.ijms.2021.116657

[58] Reddy GVR, Reddy BV, Haque SW, Gautam HD, Kumar P, Kumar AP, et al. Development and validation of a stability-indicating uplc method for rosuvastatin and its related impurities in pharmaceutical dosage forms. Quimica Nova. 2011;**34**(2):250-255. DOI: 10.1590/S0100-40422011000200015

[59] Jagadabi V, Nagendra Kumar PV, Mahesh K, Pamidi S, Ramaprasad LA, Nagaraju D. A stability-indicating UPLC method for the determination of potential impurities and its mass by a new QDa mass detector in Daclatasvir drug used to treat hepatitis C infection. Journal of Chromatographic Science. 2019;**57**(1):44-53. DOI: 10.1093/chromsci/bmy079

[60] Jane BK, M. Development and validation of stability- indicating UPLC method for the determination of gliclazide and its impurities in pharmaceutical dosage forms. Research Journal of Pharmacy and Technology. 2021;7(95):1-10. DOI: 10.1186/s43094-021-00248-w

[61] Prabhu RC, Maruthapillai A. New RP-UPLC method development using QbD approach for determination of mebendazole, quinfamide, its impurities and antioxidants in mebendazole and quinfamide fixed dose combinations (FDC). Mater Today Proc. 2020;**40**:S120-S126. DOI: 10.1016/j.matpr.2020.04.243

[62] Katakam LNR, Dongala T, Ettaboina SK. Quality by design with design of experiments approach for development of a stability-indicating LC method for enzalutamide and its impurities in soft gel dosage formulation. Biomedical Chromatography. 2021;**35**(5):1-11. DOI: 10.1002/bmc.5062

[63] King S, Stoffolano PJ, Robinson E, Eichhold TE, Hoke SH, Baker TR, et al. The evaluation and application of UPLC for the rapid analysis of dose formulations. LC-GC North Am. 2005;**23**(5):36-39

[64] Ashraf SA, Nazir S, Adnan M, Azad ZRAA. UPLC-MS: An emerging novel technology and its application in food safety. In: Srivastva AN, editor. Analytical Chemistry-Advancement, Perspectives and Applications. London, UK: IntechOpen; 2020. DOI: 10.5772/intechopen.92455

[65] (ICMSF) TIC on MS for F. International commission on microbiological specifications for foods. 2006.

[66] Ouakhssase A, Fatini N, Ait AE. A facile extraction method followed by UPLC-MS/MS for the analysis of aflatoxins and ochratoxin A in raw coffee beans. Food Addit Contam-Part A Chem Anal Control Expo Risk Assess. 2021;**38**(9):1551-1560. DOI: 10.1080/19440049.2021.1925165

[67] Calbiani F, Careri M, Elviri L, Mangia A, Pistara L, Zagnoni I. Development and in-house validation of a liquid chromatography – electrospray tandem mass spectrometry method for the simultaneous determination of Sudan I, Sudan II, Sudan III, and Sudan IV in hot chili products. Journal of Chromatography. A. 2004;**1042**:123-130. DOI: 10.1016/j.chroma.2004.05.027

[68] Romera D, Mateo EM, Mateo-Castro R, Gómez JV, Gimeno-Adelantado JV, Jiménez M. Determination of multiple mycotoxins in feedstuffs by combined use of UPLC–MS/MS and UPLC–QTOF–MS. Food Chemistry. 2018;**267**:140-148. DOI: 10.1016/j.foodchem.2017.11.040

[69] Vilkickyte G, Motiekaityte V, Vainoriene R, Liaudanskas M, Raudone L. Development, validation, and application of UPLC-PDA method for anthocyanins profiling in Vaccinium L. berries. Journal of Berry Research. 2021;**11**(4):583-599. DOI: 10.3233/jbr-200658

[70] Li L, Wang J, Li M, Yang Y, Wang Z, Miao J, et al. Detection of the adulteration of camel milk powder with cow milk by ultra-high performance liquid chromatography (UPLC). International Dairy Journal. 2021;**121**:105117. DOI: 10.1016/j.idairyj.2021.105117

[71] Liu X, Yang JL, Li JH, Li XL, Li J, Lu XY, et al. Analysis of water-soluble azo dyes in soft drinks by high resolution UPLC-MS. Food Addit Contam-Part A Chem Anal Control Expo Risk Assess. 2011;**28**(10):1315-1323. DOI: 10.1080/19440049.2011.604795

[72] Ren XN, Yin SA, Yang ZY, Yang XG, Shao B, Ren YP, et al. Application of UPLC-MS/MS method for analyzing B-vitamins in human Milk. Biomedical and Environmental Sciences. 2015;**28**(10):738-750. DOI: 10.3967/bes2015.104

[73] Vardali SC, Samanidou VF, Kotzamanis YP. Development and

validation of an ultra performance liquid chromatography-quadrupole time of flight-mass spectrometry (in MSE mode) method for the quantitative determination of 20 antimicrobial residues in edible muscle tissue of European sea bass. Journal of Chromatography. A. 2018;**1575**:40-48. DOI: 10.1016/j.chroma.2018.09.017

[74] Zheng W, Fang ZL, Yu ZK, Yang WX, Qun XF. Determination of Tetracyclines and their Epimers in agricultural soil fertilized with swine manure by ultra-high-performance liquid chromatography tandem mass spectrometry. Journal of Integrative Agriculture. 2012;**11**(7):1189-1198. DOI: 10.1016/S2095-3119(12)60114-2

[75] Claassen C, Ebel E, Kuballa J, Rohn S. Impacts of fungicide treatment and conventional fertilization management on the potato metabolome (Solanum tuberosum L.) evaluated with UPLC-IMS-QToF. Journal of Agricultural and Food Chemistry. 2019;**67**(41):11542-11552. DOI: 10.1021/acs.jafc.9b03911

[76] Isaak CK, Petkau JC, Karmin O, Ominski K, Rodriguez-Lecompte JC, YLS. Seasonal variations in phenolic compounds and antioxidant capacity of Cornus stolonifera plant material: Applications in agriculture. Canadian Journal of Plant Science. 2013;**93**(4): 725-734. DOI: 10.4141/cjps2012-310

[77] Duan S. Study on analytical detection technology of pesticide residue in agricultural product food safety. IOP Conference Series: Earth and Environmental Science. 2021;**769**:022015

[78] Morcol TB, Wysocki K, Sankaran RP, Matthews PD, Kennelly EJ. UPLC-QTof-MSE metabolomics reveals changes in leaf primary and secondary metabolism of hop (Humulus lupulus L.) plants under drought stress. Journal of Agricultural and Food Chemistry. 2020; **68**(49):14698-14708. DOI: 10.1021/acs.jafc.0c05987

[79] Reddy TSK, Balammal G, Kumar a S. Ultra performance liquid chromatography: An introduction and review. INTERNATIONAL JOURNAL OF PHARMACEUTICAL RESEARCH AND ANALYSIS. 2012;**2**(1):24-31

[80] Jena BR, Babu SM, Pradhan DP, Swain S. UPLC analytical method development and validation for the simultaneous estimation of paracetamol and caffeine capsules dosages form. Pharmaceutical Regulatory Affairs. 2017; **6**(1):1-9. DOI: 10.4172/2167-7689.1000186

[81] Waters Corporation. What is the difference bettwen UHPLC and UPLC. 2014. Available from: https://www.chromatographytoday.com/news/hplc-uhplc/31/breaking-news/what-is-the-difference-between-uhplc-and-uplc/30754 [Accessed: December, 2021].

[82] Jerman Klen T, Golc Wondra A, Vrhovšek U, Mozetič VB. Phenolic profiling of olives and olive oil process-derived matrices using UPLC-DAD-ESI-QTOF-HRMS analysis. Journal of Agricultural and Food Chemistry. 2015; **63**(15):3859-3872. DOI: 10.1021/jf506345q

[83] Saxena SK, Rangasamy R, Krishnan AA, Singh DP, Uke SP, Malekadi PK, et al. Simultaneous determination of multi-residue and multi-class antibiotics in aquaculture shrimps by UPLC-MS/MS. Food Chemistry. 2018;**260**:336-343. DOI: 10.1016/j.foodchem.2018.04.018

Chapter 2

Turbulent Flow Chromatography: A Unique Two-Dimensional Liquid Chromatography

Francesca Di Gaudio, Annamaria Cucina and Sergio Indelicato

Abstract

Among 2D-LC techniques, a particular approach is commercialized by Thermo Fisher Scientific that may enable the direct introduction of biological samples into an online automated extraction system without any additional pre-treatment: the TurboFlow technology. It combines chemical and size exclusion capability of chromatography columns packed with porous particles in which a turbulent solvent flow is able to separate smaller molecules from larger ones (e.g. proteins). Once extracted, the small molecules can also be transferred to an analytical column for improving separation prior to detection. This is done through a unique plumbing and customized valve-switching arrangement that allows the focusing of molecules onto the second column. This enables a very efficient chromatographic separation. The use of the TurboFlow not only eliminates extensive sample preparation, thus reducing inter-operator variability and matrix effects, but also increases the capacity for high-throughput analyses due to a unique multiplexing technology, in which multiple LC channels are connected to a single detector.

Keywords: 2D-LC, turbulent flow chromatography, TurboFlow, multiplexing, extraction, purification

1. Introduction

High-performance liquid chromatography (HPLC) is a powerful technique for the separation of compounds. However, one-dimensional liquid chromatography (1D-LC) cannot easily handle complex matrices and complex mixtures of analytes. The matrix effect is relevant when dealing with biological matrices, such as urine, saliva, serum, and whole blood in liquid chromatography coupled with mass spectrometry (LC–MS). This effect is mainly due to co-elution of endogenous compounds, such as proteins, lipids, sugars, or salts [1, 2]. In fact, early approaches in LC–MS, tending to simplify sample preparation methods, such as "dilute and shoot" and quick chromatographic analysis time, soon revealed that by not removing the matrix components, these could interfere with the ionization process in an unpredictable and inconsistent way (e.g. ion suppression).

For these reasons, a thorough and robust sample preparation is crucial for quantitative analysis. Among others, protein precipitation (PP) methods are widely used. In these procedures, a small volume of sample is mixed with a certain volume of protein precipitation reagent. The interaction between proteins and this reagent determines the alteration of the protein conformation and the consequent precipitation. Once the precipitate is separated, the analytes of interest, remaining in solution, are ready for analysis. Although PP protocols are inexpensive and simple, they are time-consuming and unable to purify the samples enough. More recently, modified PP approaches, such as membrane-based PP filter plates, have been proposed to overcome these limitations [3, 4].

Liquid–liquid extraction (LLE) and solid-phase extraction (SPE) are alternative sample preparation methods. The first one consists in extracting the analytes from one liquid phase to another immiscible liquid phase. Once shaken or vortex-mixed, the phases are separated, and the molecules collected for analysis. LLE presents several limitations, such as the need for large sample volume, variable recovery, and difficulty on the extraction of compounds with varying lipophilicities. A more efficient upgrade of this technique is represented by salting-out assisted LLE (SALLE), in which appropriate salts are added to the solvent to enhance the extraction [5].

SPE technique allows extraction and enrichment of analytes of interest by loading a liquid sample onto a column/cartridge/plate packed with a sorbent material. While the analytes are retained, the interferents are either eliminated during the loading phase or washed away during the washing steps. The analytes are then eluted and collected for LC–MS analysis [6]. Even if SPE has many advantages, it has limited selectivity and sensitivity because matrix constituents can be adsorbed together with the analytes, determining matrix effects. Furthermore, SPE cartridges are usually single use. More recently, several approaches have been developed to improve SPE technique, such as dispersive solid-phase extraction, solid-phase micro-extraction, and stir bar sorptive extraction [7]. Furthermore, SPE has been automated in online solid phase extraction, coupled with HPLC [8]. This technique can thus be considered as an online two-dimensional liquid chromatography (2D-LC).

The online 2-D approaches are faster and more reproducible, but they require specific interfaces and more complex operative modes.

There are two main approaches to the online 2D-LC: heart-cutting and comprehensive [9]. The first one enables the re-injection of a definite number of multicomponent effluent fractions from a primary to a secondary column, while the second one determines the separation of the entire sample in both dimensions [10]. When this technique is employed, a combination of different chromatographic separation methods is usually utilized. Frequent combinations are size exclusion chromatography (SEC) with reverse phase liquid chromatography (RPLC), RPLC with hydrophilic interaction liquid chromatography (HILIC), or normal-phase liquid chromatography with RPLC [11].

Regardless of the operative mode, 2D-LC presents some disadvantages. In fact, the separation typically takes longer than in 1D-LC, and the detection sensitivity may decrease because of the dilution or loss of the sample during the two separations. In addition, considering that the 1D effluent is the injection solvent of the second dimension, incompatibility issues may be raised. Incompatibility could be determined by partial or complete immiscibility of the two mobile phases, difference in solvent strength, but also by the excessive difference in viscosity, that could result in peak deformation or splitting. Besides, an increased instrumentation and conceptual complexity must be considered. Active-modulation techniques, such as active solvent modulation (ASM) and stationary-phase assisted modulation (SPAM), could surmount these limitations. However, a constraint to the ASM approach is the time needed to displace the fractions of effluent from the sampling loop to the second column. The required extra-time can be significant for short 2D cycles [12]. SPAM separation methods of analytes with completely different chemical properties could lead to incomplete recovery and discrimination effects, due to a partial trapping of all analytes with the chosen dilution solvent. Moreover, the trapping columns of SPAM could reduce the robustness of the system [13].

Turbulent flow chromatography (TFC) overcomes the limitations of 2D-LC by removing potential interferences in a fast and efficient manner. In addition, the application of staggered, parallel methods ("multiplexing") maximizes the use of detector time, saving time and solvents. Furthermore, the extraction columns are re-usable for hundred injections [14].

In this chapter, theory and hardware aspects of the TFC, with particular reference to the TurboFlow system, will be discussed. Moreover, method development and applications will be brought up.

2. Theory of turbulent flow chromatography

In the ideal chromatographic process, the analytes form narrow bands while moving along the column. The band or peak narrowness is a measure of the efficiency of the separation procedure that can be quantified by the number of theoretical plates, *N*. These latter are hypothetical zones or stages in which two phases, the stationary phase, and the liquid mobile phase in the case of HPLC, establish an equilibrium with each other. If the peak shape is Gaussian, the theoretical plate number *N* can be approximated according to Eq. (1).

$$N = 5.54 * \left(\frac{t(r)}{W(1/2)}\right)^2 \tag{1}$$

where $t(r)$ is the retention time of a compound, and $W(1/2)$ is the peak width measured at half of the peak height. As described in the Eq. (2), the theoretical plate number *N* depends on the ratio between the column length and the height equivalent of a theoretical plate, *h*.

$$N = \frac{L}{h * d(p)} \tag{2}$$

where *L* is the column length, *N* is the number of theoretical plates, and $d(p)$ *is* the particle diameter. The smallest value of *h* corresponds to the highest efficiency. Thus, *h* is a more effective way to compare efficiency of the chromatographic processes.

Column efficiency, linked to band broadening processes, in traditional HPLC was first described by van Deemter and colleagues [15]. Van Deemter described these mechanisms with the Eq. (3).

$$h = A + \frac{B}{v} + Cv \tag{3}$$

where *A*, *B*, and *C* are constants, and v is the average linear velocity (cm/s). The optimum value of *h* will be obtained when

$$\frac{dh}{dv} = 0$$

so that

$$v = \sqrt{\frac{B}{C}}$$

The term A is defined as the tortuosity factor, and it is affected by the size and the distribution of the stationary phase particles. The term B depends on the diffusion of the molecules in the longitudinal direction, linked to the motion of solute molecules in the mobile phase. This coefficient is significant at low flowrates. The term C refers to the resistance to the mass transfer of the analytes through the two phases and their diffusion on the surface. **Figure 1** presents a simplified van Deemter plot, which allows us to identify the optimum flowrate, corresponding to the minimum value of h.

In 1966, Pretorius and Smuts firstly observed that, using open tubular columns, it is possible to improve speed and efficiency of mass transfer with turbulent rather than laminar flow [16]. Turbulent flow in fact, dominated by rotational motion, reduces band-broadening, presenting a profile flatter than laminar flow (**Figure 2**).

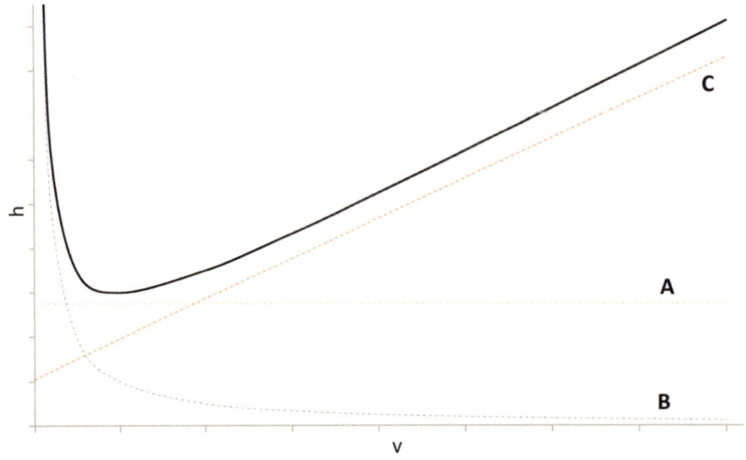

Figure 1.
Simplified van Deemeter plot with A, B and C components.

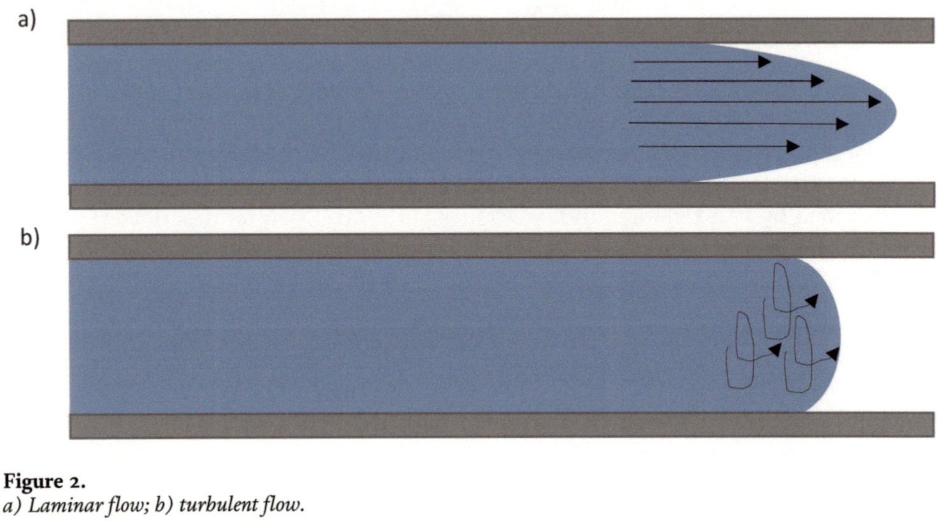

Figure 2.
a) Laminar flow; b) turbulent flow.

Physicist Osborne Reynolds found that the flow of a fluid through a straight and smooth tube transitions from laminar to turbulent as the momentum of the fluid becomes around two thousand times greater than its resistance to flow. The momentum, which is a certain amount of mass moving at a certain velocity, takes into account the fluid density and the diameter of the tube. It is possible to increase the fluid momentum by increasing its velocity, or by increasing the diameter of the tube, or both. The resistance to flow is expressed as the absolute or dynamic viscosity of the fluid [17].

The Reynolds number, Re, is a dimensionless parameter, defined as the ratio of the inertial to viscous forces present in a fluid system (Eq. 4).

$$\Re = \frac{\mu 0 * I}{\eta} \tag{4}$$

where μ_0 is the mean linear velocity of the fluid, and η is the dynamic viscosity of the mobile phase. In a packed bed such as an HPLC system, the characteristic length scale I is related to particle diameter and external porosity.

The transition from laminar to turbulent flow occurs as Re overcomes some reference values. For a very straight and smooth cylinder the transition from laminar to turbulent flow occurs at a Reynolds number of 2400, while for a packed bed of uniform spheres, turbulence occurs between 1 and 10 Re [18]. Even if turbulent flow presents the advantage of better mass transfer, the backpressures required to obtain this flow would be excessively high, particularly in terms of pumping systems and particle architecture. In the 1990s, it was observed that using large irregularly shaped particles (50–100 μm), the backpressures were sufficiently low to generate turbulent flow and to improve column efficiency at increasing flowrate thanks to large interstitial spaces between particles (around 100 Å). In addition, it was observed that larger molecules in solution diffuse more slowly than smaller ones. Thus, larger molecules do not interact with the stationary phase, while smaller molecules have time to diffuse in and out of the pores and to interact with the stationary phase depending on their affinity. In 1997, these packed columns (now called TurboFlow™ columns) were patented (US patent no. 5,919,368) [19]. Although the mechanism is not fully understood, TurboFlow has demonstrated to be a very efficient method to separate large molecules, such as matrix components (e.g. proteins and lipids), from small molecules [14].

2.1 Column chemistries

TurboFlow™ columns are available in a wide variety of packing materials to enable different applications. The selection of the appropriate column must be based on the polarity of the analytes and the mobile phases needed to solubilize them. Initially, these columns presented an internal diameter of 1 mm and required a flowrate of 4–5 mL/min. Recent columns have an internal diameter of 0.5 mm to reduce solvent consumption and require a flowrate around 1.5–2 mL/min. It is possible to group TurboFlow™ columns in silica-based and polymer-based. Among silica-based columns, suitable for pH ranging from 2 to 9, packing includes Cyclone C18, Cyclone C8, and Cyclone C2 modified columns. Furthermore, Cyclone Fluro and Cyclone C18-P columns are also available. Styrene/divinylbenzene copolymers-based columns, suitable for pH ranging from 1 to 13, are also available in different chemistries: Cyclone, Cyclone-P, Cyclone mixed-mode with anion—(Cyclone MAX), cation—(Cyclone MCX), and mixed anion–cation-exchange phases (Cyclone MCX-2). Details about column applications are described in **Table 1** [20].

Column	Application
Silica-based columns	
Cyclone C18	For non-polar and moderately polar solutes (f.e. pharmaceuticals, fatty acids)
Cyclone C18-P	For polar and non-polar solutes (f.e. polar pharmaceuticals, metabolites)
Cyclone C8	For solutes excessively retained by C18 (f.e. pharmaceuticals, metabolites)
Cyclone Fluro	For highly-lipophilic solutes and perfluorinated compounds
Cyclone C2	For extremely non-polar solutes (f.e. extremely non-polar multi-functional pharmaceuticals)
Polymer-based columns	
Cyclone	For non-polar analytes in complex matrices and silanophilic compounds
Cyclone-P	For non-polar and moderately polar compounds (f.e. steroids)
Cyclone MAX	For polar solutes and weakly acidic compounds (f.e. polar pharmaceuticals, metabolites, antibiotics)
Cyclone MCX	For polar solutes and weakly basic compounds (f.e. polar pharmaceuticals, metabolites, drugs of abuse)
Cyclone MCX-2	For polar solutes and weakly basic compounds (f.e. polar pharmaceuticals, metabolites, metanephrines, melamine)

Table 1.
Silica-based and polymer-based column applications.

3. Hardware

The turbulent flow chromatography, whose main peculiarity lies in the particular characteristics of its columns (described above), requires a specific hardware platform capable of exploiting the potential that these can express. The only instruments on the market designed to take full advantage of TurboFlow columns are marketed by Thermo Fisher Scientific as Transcend TLX systems.

These systems can be used as traditional HPLC, injecting the sample directly onto the analytical column, or for 2D-LC. In order to handle a turbulent flow chromatography and eventually to couple it with a second analytical column, a complex and unique hardware, which presents at least two pump systems and a valve interface module, is required. In addition, the systems can include multi-channel technologies. Depending on its configuration, each system can include one, two, or four pairs of pumps that enable the system to work either in single or 2D-LC mode over one, two, or four separate channels. When a sample is extracted/purified, the system uses one pump, named loading pump, to carry the sample into the TurboFlow column, and a second pump, the eluting pump, to resolve the extracts over a traditional analytical column. The overall plumbing of the system is intricate since it can include up to four pairs of valves each lodging a different couple of columns (the TurboFlow column in valve A and the analytical one in valve B) and, eventually, a special extraction loop, together with other valves for channel selection and for flow diversion to the detector or to the waste (**Figure 3**). Moreover, the system controls up to four couples of injection ports (for the online extraction and for the traditional elution) using one or two autosampler arms. Some interesting accessories can complete and further complexify the configuration, as for example a multicolumn compartment module (MCM) that allows to install simultaneously up to six extraction and six analytical

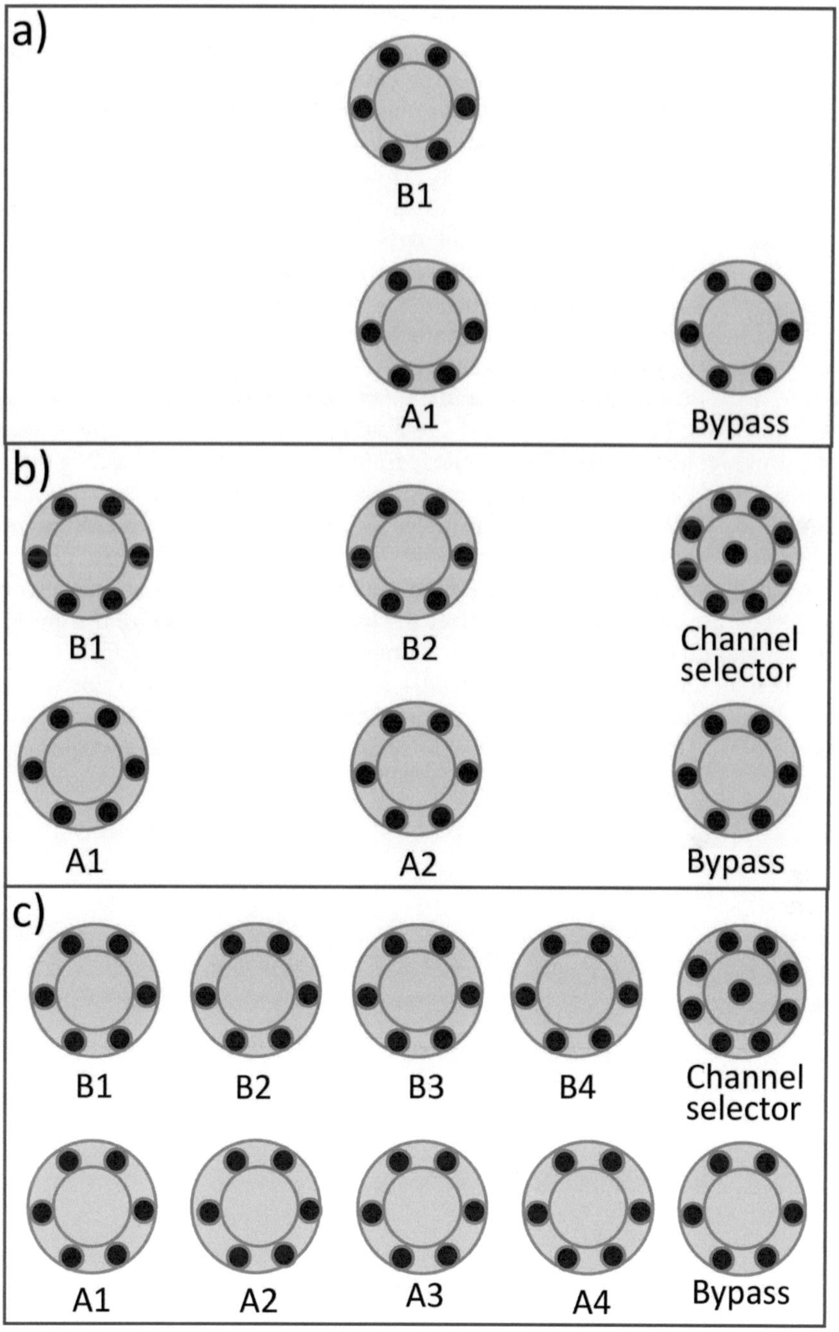

Figure 3.
Valve interface module of a) single channel, b) double channel, c) four channel systems. The "bypass" valve diverts the flow to the waste or to the detector.

columns (or eleven columns of the same type) and to select the column to be used via software, without the need to plug and unplug them. Considering the complexity of Transcend systems, a dedicated software (Aria MX™) simplifies their use by controlling and coordinating the valve switching processes and the multiplexing capabilities.

Using Transcend systems for online extraction, two modes of operation are possible: Quick Elute Mode and Focus Mode (**Figures 4** and **5**). These differ in complexity, field of application and plumbing, being the Quick Elute mode simpler and faster.

Quick Elute is generally used for single compound quantification and/or when speed is the priority. In this case, in which only a TurboFlow column is needed, chromatographic separation is performed, while compounds are eluted from the column. This operating mode presents the advantages of being fast and simple to be set even when analysing complex matrices. Furthermore, it requires minimal method development. However, this separation presents a limited resolution, because of the large irregularly shaped particles of TurboFlow columns, so possible isobaric interferences should be taken into account. Nonetheless, it is possible to add an analytical column between the valve B and the detector for a rapid 2-D separation, similar to an online SPE, with the advantages that a turbulent flow gives to the extraction phase.

The Quick Elute Mode consists of the following steps: loading, eluting, column cleaning, and reconditioning. The sample is firstly injected into the loading pump mobile phase flow. The analytes are brought to and retained by the column, installed on the A valve, while the matrix macromolecules, not retained by the column, are washed to the waste connected in the B valve. If the analytes are not retained, they will be washed away in this step, if they are too retained, they will be washed out during the column cleaning process. After the loading step and before the elution, the TurboFlow column can be rinsed with a stronger mobile phase to purify the sample.

By switching the B valve, the mobile phase from the eluting pump flows through the TurboFlow column, the analytes are eluted to the detector or, if the case, to an analytical column. It is possible to elute the analytes off the TurboFlow column using

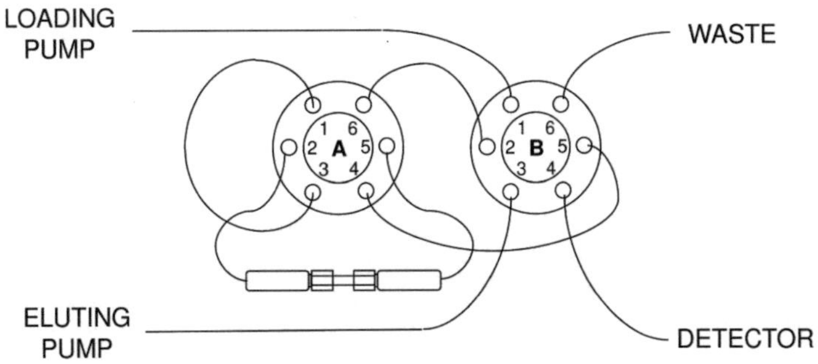

Figure 4.
Quick elute valve arrangement without analytical column.

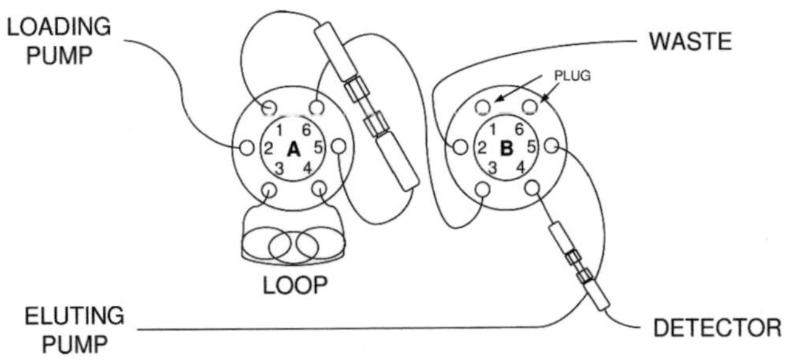

Figure 5.
Focus mode valve arrangement.

the same flow direction as the loading step (forward flush), or in the opposite direction (back flush), by switching the A valve. After the elution step, the extraction column is washed by the loading pump flow with an appropriate organic solvent. The re-conditioning presents the same parameters of the loading step, but for a longer time to better equilibrate the system.

When the quantification of multiple compounds is requested, Focus Mode is indispensable. It involves a TurboFlow and an analytical column and requires more extensive method development. In addition to the Quick Elute mode setup, the Focus Mode requires an elution loop in the A valve for the transfer step and a customized tee-piece rotor seal in the B valve (**Figure 6**).

The Focus Mode consists of six steps: loading, sample washing, transfer, system washing, loop-filling, and reconditioning (**Figure 7**) [20]. In the loading step, samples are injected into the TurboFlow column at high flow rates so that macromolecules of the matrix, salts and ionic compounds, not retained by the column, flow to waste (**Figure 7a**). Once loaded, the sample is cleaned from eventual contaminants using an appropriate mobile phase (**Figure 7b**). During these two steps, the mobile phase of the eluting component flushes to the analytical column connected to the detector.

During the transfer step, the valves A and B are switched, the optimized loop content is pushed by the mobile phase into the TurboFlow column, and the analytes are eluted off and transferred to the analytical column (**Figure 7c**). In this step, the transfer flow is diluted, via the tee-piece rotor seal, in the B valve, with the mobile phase coming from the eluting pump, in order to focus the molecules of interest to the head of the analytical column. These first three steps typically take around 2 minutes.

When the transfer is completed, the B valve is switched, the loading and eluting flows are again separated and, while the analytes are resolved on the analytical column and eluted to the detector, the TurboFlow column and the extraction components are washed to avoid carryover (**Figure 7d**). At this moment, the loop is filled with the proper transfer solution to be used in the following injection (**Figure 7e**). In the final step, the elution loop is isolated, and the columns are re-equilibrated to the conditions of the loading step (**Figure 7f**).

3.1 Operations in multichannel systems

In a traditional LC run, analytes of interest are usually eluted in a small portion of the entire chromatogram. It means that, during the pre-injection phase (needle and injector wash, sample withdrawn, etc.), but also during the first and last minutes of a run, the detector acquires useless data or the flow is diverted to waste. TurboFlow is multichannel systems, in which each "channel" is an independent HPLC equipment with its own pump(s) and injector port(s). These channels are connected to one single detector, and the systems are able to maximize the productivity, reducing the idle

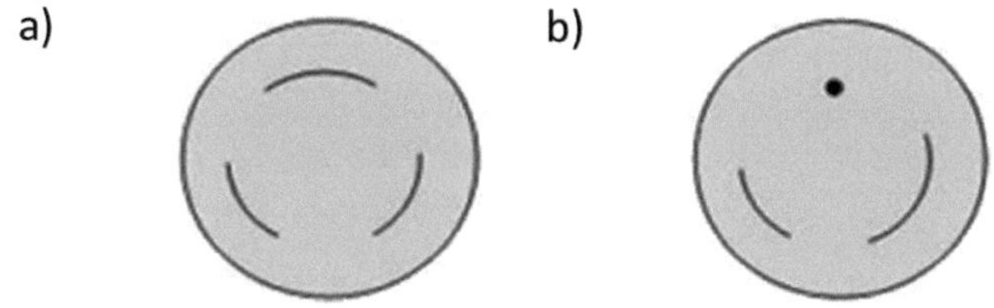

Figure 6.
Rotor seals in a) quick elute mode, b) focus mode.

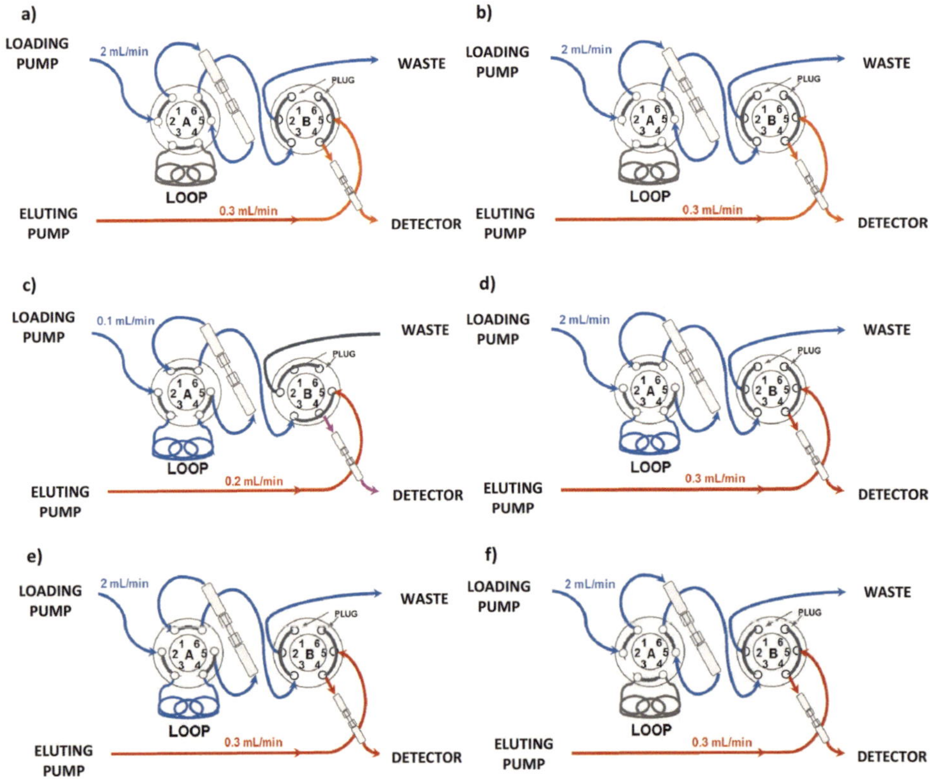

Figure 7.
a) Loading; b) sample washing; c) transfer; d) system washing; e) loop filling; f) reconditioning.

time of the detector. Thanks to the multiplexing, in fact, TurboFlow staggers injections, by overlapping the channels to optimize the time of the acquisition window. In particular, two- or four-channel systems are currently available (**Figure 8**). Aria MX™ software is able to calculate automatically the timing of injections based on the relative duration of acquisition windows in comparison with the duration of the entire chromatogram and the pre- and post-injection operations of the auto-sampler. On each of

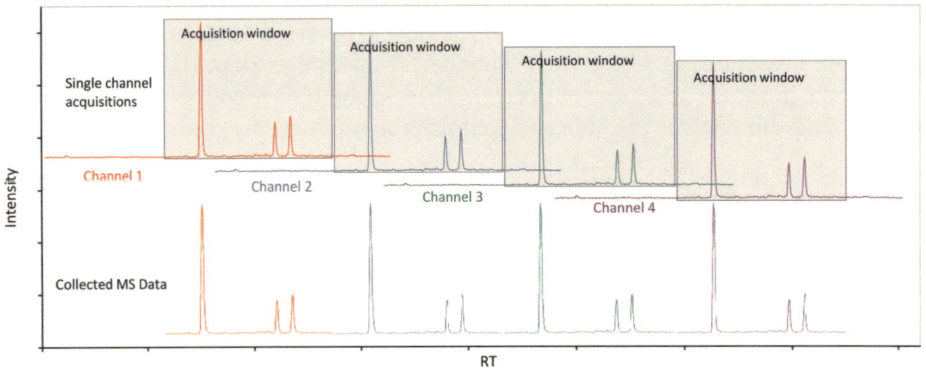

Figure 8.
Representation of staggered injections in a multiplexing system.

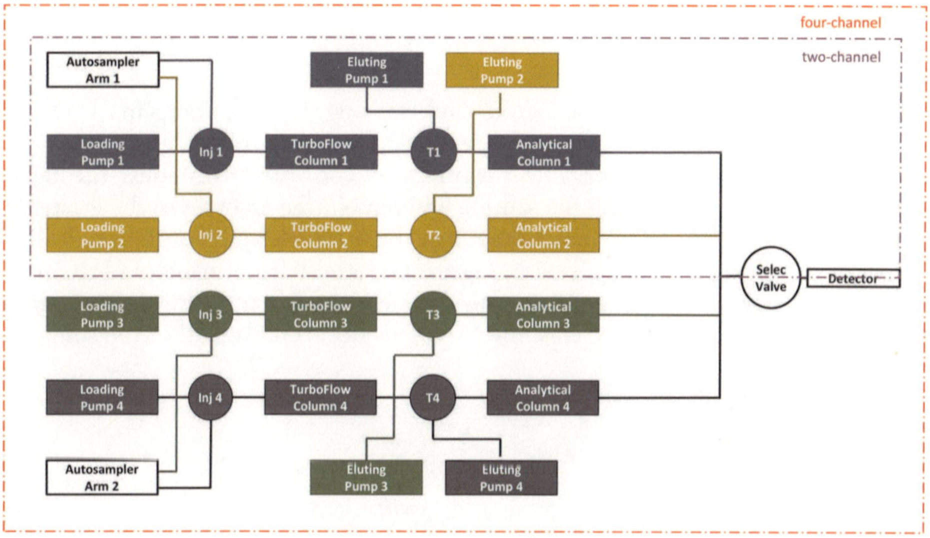

Figure 9.
Diagram showing the setup of a two- (squared in purple) and a four-channel (squared in red) multiplexing.

the different channels, the system is able to handle runs with the same or different instrumental method, maximizing the productivity up to four times when using a four-channel system. However, the hardware complexity is also significantly increased (**Figure 9**).

4. Development of a focus mode method

The two approaches described above require different method development. However, considering the complexity of the Focus Mode, in the present paragraph only its method development will be explained.

A Focus Mode involves at least two columns, a TurboFlow and an analytical column. Two mobile phases (appropriately chosen) for the analytical gradient elution and at least two mobile phases for the extraction component for loading, transfer and washing steps are also requested. It is generally suggested to start with the choice of the ideal analytical column and elution conditions, able to chromatographically resolve the analytes. Then, considering the type of chromatography for the extraction process (i.e. reverse phase, ionic exchange, etc.) intended to perform, a TurboFlow column has to be chosen, compatible with the chemical characteristics of the analytes, the mobile phases, and the column selected for the analytical separation. Considering the possible combinations of couples of columns, a multiple column module (MCM) can help in this evaluation. Once the separation conditions are set, loading, transfer and washing conditions have to be optimized. This process that can be facilitated by the Aria software, consists of studying the analytes chromatographic behaviour when mobile phase composition, duration, and flowrate are changed during the different steps. In fact, one of the software useful features allows one to test different conditions for a specific method parameter, defining it as "variable" and programming its values in the acquisition sequence instead of writing one method per condition that has to be tested. When optimizing these parameters, the analytical column has to be removed. A T-union, which splits the flow of the eluting mobile phase between detector and waste, is connected to detect the analytes elution at all steps in the method (**Figures** 7 and **10**).

During the optimization of the loading and washing step conditions in a classic TurboFlow method, in which reverse phase chromatography for analytes extraction is used, different percentages of organic solvent have to be tested to establish the highest % of organic phase able to wash the sample without eluting analytes to the waste. In the wash step all interferences more polar than the molecules of interest are ideally washed out. In the example reported in **Figure 11**, tetrahydrocannabinol (THC) was loaded in a Cyclone-P column in 100% aqueous phase and an increasing percentage of organic phase was tested for the washing step. The ideal condition of the washing step

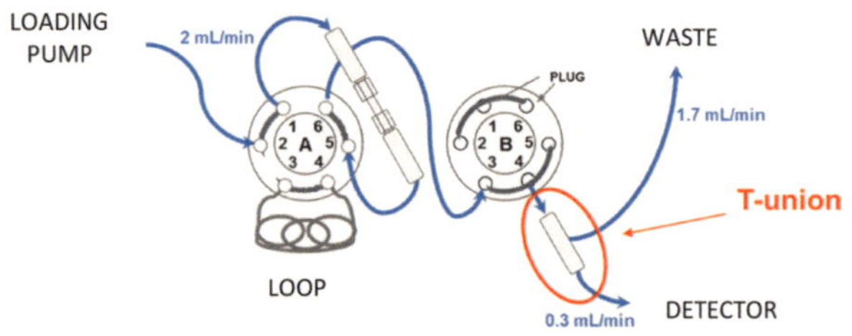

Figure 10.
Method development setup: The analytical column is replaced by a T-union diverting only the proper flow from the TX column to the MS detector.

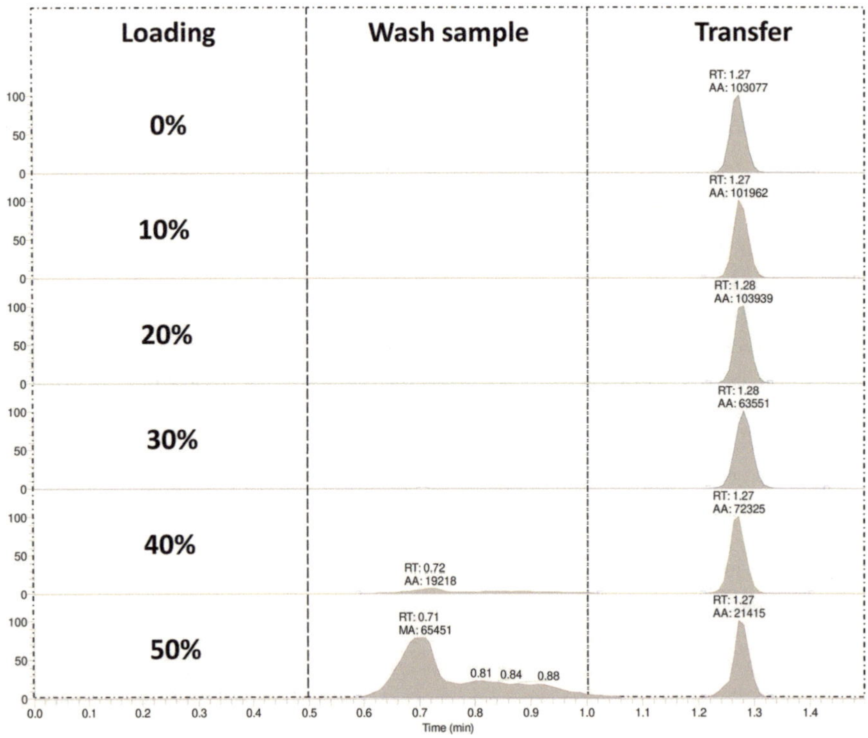

Figure 11.
Optimization of loading conditions for the analysis of THC in a cyclone-P column. The percentages refer to the organic phase.

was at 30% of B (**Figure 11**). Results with a higher organic content (40% and 50% in **Figure 11**) showed peaks appearing during the wash step, therefore demonstrating that the analytes are no longer retained on the TurboFlow column. Some loss during the loading/washing steps might be accepted as long as this is not impacting the required limits of quantification, since this will make the sample cleaning more efficient.

In the Transfer step, the analytes elute from the TurboFlow column and are transferred to the analytical column by the contents of the transfer loop. The solvent strength of the loop content is diluted before entering the analytical column by the eluting pump flow. The goal in this step is to be able to quantitatively elute the analytes from the TurboFlow column and to focus them on the head of the analytical column while reducing as much as possible the hydrophobic interferences from the sample matrix (matrix effect). So, three development stages are required: optimization of the transfer loop content, review of the transfer time using different transfer flowrates, and finally optimization of the transfer step dilution ratio.

Regarding the transfer loop content, the goal at this stage is to determine the minimum amount of organic solvent that quantitatively transfers all the target analytes to the analytical column, without eluting more hydrophobic compounds. The lower the content of organic solvent in the loop, the higher the retention of late eluting components (e.g. phospholipids) in the TurboFlow column (then washed to waste during the wash step), reducing matrix effect during the chromatographic elution.

If the solvent strength of the loop is too low, the analytes remain retained in the TurboFlow column and will be washed to waste during the washing step. On the other hand, if the solvent strength is too high, focusing of the analytes on top of the analytical column might be affected, leading to chromatographic peaks distortion or lack of retention. Moreover, the purification process would be less efficient. During the method development, the optimization of the loop content is performed without the analytical column.

In the example of the method development for the analysis of THC in the Cyclone-P column, decreasing percentages of organic phase were tested, and the ideal one was identified as 50% (**Figure 12**). Indeed, for lower content of organic solvent in the loop, part of the analyte was not transferred and was washed out in the washing step.

It is advisable to evaluate the transfer step time for different loading pump flowrates in preparation for the next step of the method development, the dilution ratio. If the flowrate of the loading pump during the transfer step is high, the method will be faster. However, the dilution from the eluting pump will be lower (taking into account that the sum of loading plus eluting flow cannot exceed the flowrate suitable for the analytical column), and hence, it will be more difficult to focus the target analytes on the head of the analytical column. This evaluation step consists in recording the time it takes for the analytes to reach the detector as the eluting flow varies in the transfer step. For the transfer time evaluation, there are no right or wrong values, but the information about the transfer that will be used in the final method, once the dilution ratio will be established, is just collected.

After the installation of the analytical column, the dilution ratio can be assessed, keeping the goal to focus the extracted analytes on the analytical column. If the organic concentration of the mobile phase in the transfer step is too high when it enters the analytical column, the analytes move through the analytical column rather than focus on it. In order to reduce this effect, the eluting pump dilutes the organic mobile phase from the loop with aqueous mobile phase reducing the solvent strength of the combined flow into the analytical column. The final solvent strength is

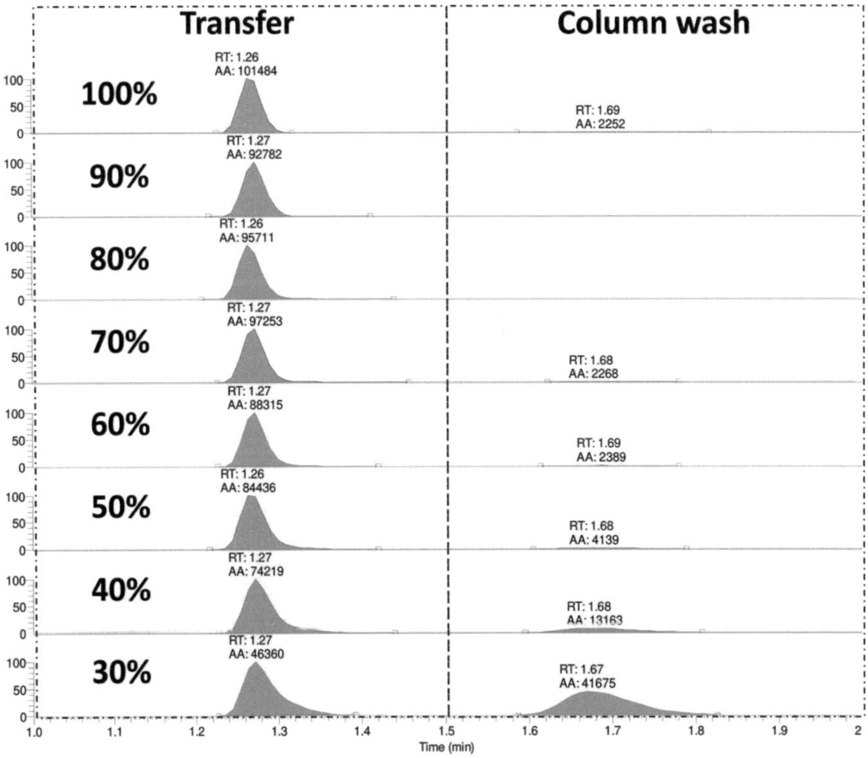

Figure 12.
Optimization of loop fill content for the analysis of THC in a cyclone-P column. The percentages refer to the organic phase.

influenced by the organic concentration of the loop contents and of the eluting pump flow (already optimized in the previous steps) and by the relative ratio between the eluting and loading pump flowrates (Eq. 5).

$$final\%organicsolvent = \left(\frac{LP}{T}\right) * OL + \left(\frac{EP}{T}\right) * OEP \quad (5)$$

where T is the total flowrate, the combined flowrate of the loading and eluting pumps during the Transfer step; LP is the loading pump flowrate during the transfer step; EP is the eluting pump flowrate during the transfer step; OL is the organic content percentage in the loop, and OEP is the organic percentage in the eluting pump flow.

Thus, the transfer dilution ratio has to be optimized by testing the effect of different combinations of loading and eluting flowrates on chromatography, taking into consideration that the total flowrate should match the flowrate used for the analytical column.

Peak fronting, poor resolution, and breakthrough in the final chromatography could be observed if solvent strength is too high during transfer and should be fixed trying to reduce the overall organic content during the transfer step.

Once the method is optimized, the maximum injection volume, which follows the same rules of a regular liquid chromatography method, has to be evaluated. The injectable sample volume will be higher when its composition is more similar to the

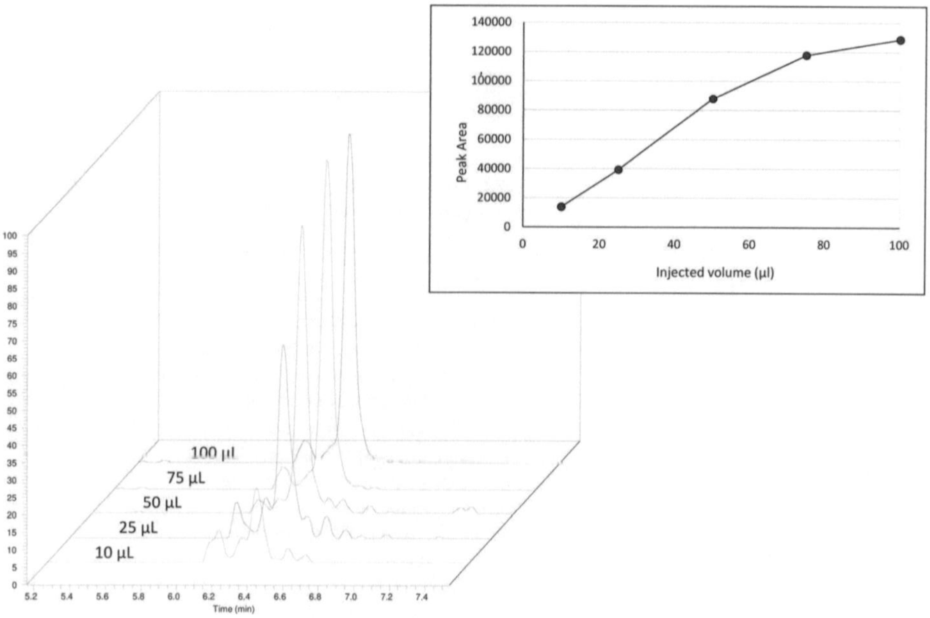

Figure 13.
Maximum injection volume identification.

initial mobile phase composition. However, it has to be considered that turbulent flow, compared to laminar flow, allows the use of a higher percentage of organic solvent. Increasing injection volumes of an extracted matrix sample, correlated to proportional increase of peak area, should be tested. The maximum injection volume corresponds to the volume that causes a flattened out peak area. The example in **Figure 13** shows that the maximum injection volume is 75 μL. Nonetheless, thanks to high carbon load values, the columns capabilities usually can handle volumes of 100 μL, significantly improving the overall method sensitivity, compared with injection volumes of conventional methods (5–10 μL).

5. Applications

Successful applications of the turbulent flow technique have been reported in different fields, such as therapeutic drug monitoring and environmental analysis, applied to various matrices, such as urine, plasma, but also food materials and water (**Table 2**).

Most of the published papers are focused on clinic applications. In this field, sample pretreatment, separation, and detection need to be more integrated in order to make mass spectrometry routinary even for not specialized laboratories. Online multidimensional chromatography, like TurboFlow methods, combining sample preparation and analysis, facilitate sample introduction, ease-of-use, and speed, even when dealing with complex matrices. Turbulent flow chromatography allows a simplified analysis of plasma, serum, or whole blood [21–26]. Hervious and colleagues reported the development, validation, and application of LC–MS/MS coupled with TurboFlow for the quantification of irinotecan, a cytotoxic agent used for metastatic

Reference	Matrix	TurboFlow column	Analytes	Application
Michopoulos [21]	Blood plasma	Cyclone	metabonomic analysis	Clinic
Chassaing [22]	Blood plasma	Cyclone	fluconazole, UK-112,166, dofetilide, candoxatril, UK-250,300, UK-141,495, doxazosin, CP-122,288, Compound I	Clinic
Gous [23]	Blood plasma	Cyclone C18-P-XL	anticoagulants: apixaban, dabigatran, edoxaban, and rivaroxaban	Clinic
Lindner [24]	Blood serum	Cyclone	amitriptyline, atomoxetine, citalopram, clomipramine, clozapine, descitalopram, desfluoxetine, desipramine, desmirtazapine, doxepine, dulozetine, fluoxetine, fluvoxamine, imipramine, maprotiline, mianserin, mirtazapine, norclomipramine, norclozapine, nordoxepine, nortryptiline, nortrimipramine, norvenlafaxine, paroxetine, protryptiline, reboxetine, sertraline, trazodone, trimipramine, venlafaxine	Clinic
Peake [25]	Blood	Cyclone-P	cyclosporine A, tacrolimus, and sirolimus	Clinic
Couchman [26]	Blood plasma or serum	C18	antifungal drugs: fluconazole, voriconazole, posaconazole, ketoconazole, itraconazole, hydroxyitraconazole, R051012	Clinic
Hervious [27]	Blood plasma	Cyclone P	irinotecan	Clinic
Helfer [28]	Blood plasma	Cyclone and Phenyl-	mixture A: morphine, levetiracetam, amphetamine, codeine, risperidone, diphenhydramine, quetiapine, promethazine, amitriptyline, sertraline, thiopental, diazepam, diclofenac, flufenamic acid; mixture B: moxonidine, hydrochlorothiazide, molsidomine, triamterene, minoxidil. Torasemide-M (HOOC-), bisoprolol, ivabradine, torasemide, ramipril-M (deethyl-), doxazosin, verapamil, aliskiren, amlodipine, ramipril, losartan, losartan-m (HOOC-), bezafibrate, spironolactone-M (carnenone), phenprocoumon, amiodarone-m (deethyl-), amiodarone	Clinic
Couchman [26]	Blood plasma or serum	Cyclone	TKIs	Clinic
Kasper [29]	Dried bloodspots	Cyclone P	enzymes for high-throughput screening of lysosomal storage disorders	Clinic
Frederiksen [30]	Urine	Cyclone P	biphenol A and other phenols	Clinic
Lopez-Garcia [31]	Urine	Cyclone P	neonicotinoids: imidacloprid, acetamiprid, clothianidin, dinotefuran, nitenpyram, thiacloprid, thiamethoxam, acetamiprid-N-desmethyl	Toxicology

Reference	Matrix	TurboFlow column	Analytes	Application
Lopez-Garcia [32]	Urine	Cyclone P	pesticides: organophosphates, pyrethroid insecticides	Toxicology
Helfer [33]	Urine	Cyclone and Phenyl-	toxins	Toxicology
Schaefer [34]	Urine	Cyclone MAX	opiates, cocaine, amphetamines, methadone, benzodiazepines	Forensic
Ates [35]	Wheat, maize, animal feed	Cyclone MCX-2	fusarium mycotoxins	Food quality
Ates [36]	Wheat, maize, animal feed	Cyclone MCX-2	plant and fungal metabolites	Food quality
Roach [37]	Infant formula	Cyclone MCX-2	melamine	Food quality
Mottier [38]	Honey	Cyclone	fluoroquinolones	Food quality
Lopez-Gutierrez [39]	Royal belly	Cyclone P	polyphenols	Food quality
Presta [40]	Wine	C18	flavonoids	Food quality
Fan [41]	Chinese cabbage and cucumber	C18	pesticides: imidacloprid, 3-hydroxycarbofuran, acetamiprid, pirimicarb, aldicaarb, propoxur, carbofuran, carbanyl, isoprocarb, metalaxyl, carbendazim, isazofos, diflubenzuron, chlorbenzuron, phoxim	Food quality
Hollosi [42]	Wheat flour and carrot-based puree	Cyclone MCX-2	48 polar, mid- and non-polar pesticides	Food quality
Bousova [43]	Chicken meat	Cyclone P	antibiotics	Food quality
Nathanail [44]	Cereal grains	Cyclone MCX-2	deoxynivalenol and its derivative	Food quality
Asperger [45]	Water and surface	C18, Phenyl, Cyclone (tested also SPE with Oasis HLB and porous graphitized carbon–Hypercarb)	pesticides: isoproturon, diuron, chlortoluron, atrazine, simazine, terbutylazine	Environment
Koal [46]	Water and surface	C18, Phenyl, Cyclone (tested also SPE with Oasis HLB and porous	28 pesticides (triazines, phenylureas, organophosphorous species and others)	Environment

Reference	Matrix	TurboFlow column	Analytes	Application
		graphitized carbon–Hypercarb)		
Lopez-Serna [47]	Water	Cyclone, Cyclone P, Cyclone MCX, C18, C18-P XL and Cyclone MAX	58 pharmaceuticals and 19 metabolites/ transformation products	Environment

Table 2.
Some applications of turbulent flow chromatography reported in literature.

colorectal cancer treatment, and its active and inactive metabolites, SN38 and SN38-G, respectively, in plasma after protein precipitation [27]. The same approach was applied for human plasma screening of various drugs that could be identified also below the therapeutic concentration [28]. Therapeutic drug monitoring of tyrosine kinase inhibitors (TKIs) is crucial for various cancers treatment. Early analytical methods involved HPLC-UV. However, not all TKIs present a good UV absorbance. TurboFlow LC–MS/MS was successfully used for the quantification of these compounds in a single analysis. In this protocol development, a precipitation step before online extraction was proven to maximize column life-time and minimize risk of autosampler blockage [14]. Multiplexed, multi-dimensional uHPLC–MS/MS was applied not only to the analysis of human plasma, but also to the high-throughput screening of lysosomal storage disorders in newborn dried bloodspots to obtain an online sample clean-up [29]. Determination of contaminants in urine, such as Bisphenol A and other environmental phenols [30], pesticides and metabolites [31, 32] or toxins [33] have been reported. Turbulent flow chromatography finds its application also in forensic toxicology. Mueller and colleagues developed a fully automated toxicological screening system for online urine extraction and analysis [48]. In the context of driving ability diagnostic, TurboFlow methods for the quantification of opiates, cocaine, amphetamines, methadone, and benzodiazepines have been validated in urine matrices [34].

Although the applications in drug monitoring are predominant, the use of TurboFlow is also reported in food and environmental quality. In fact, food products contain analytes in low levels, and other constituents, potentially interferents, such as sugars, proteins, and pigments in significantly higher concentrations, determining the necessity of sample clean-up and/or pre-concentration. The applicability of this technique in food analysis has already been proved in the determination of Fusarium mycotoxins [35] and of plant and fungal metabolites in wheat, maize, and animal feed [36], but also in the determination of melamine in infant formula [37]. A successful method has been developed and applied by Mottier and colleagues for the determination of 16 fluoroquinolones in honey, used to treat bee's bacterial diseases [38]. This technique allowed the reduction of extraction time and elimination of interferences in complex samples such as royal jelly and the determination of polyphenols [39]. A TurboFlow method was validated for the quantification of flavonoids and resveratrol in different types of wine. The authors suggested applying the method for the quantification of flavonoids, which could be correlated to the type of pesticides and of grapes used to produce wine [40]. Online purification based on turbulent flow chromatography for the simultaneous quantification of multiple pesticides residues is reported in different matrices, such as Chinese cabbage and cucumber samples [41], grape, wheat flour, and carrot-based puree baby food [42]. Veterinary drug residues of 36 antibiotics from seven different chemical classes were identified and quantified in chicken meat, bringing the benefits of automation and cost-effectiveness [43]. The versatility in the analysis of different matrices has been confirmed in a study comparing conventional and online sample clean-up system for the determination of deoxynivalenol and its conjugated derivative, deoxynivalenol-3-glucoside, in cereal grains [44].

Recently, laboratories have begun to move towards alternative and greener methods for environmental analysis, in response to the growing awareness of more sustainable and environmentally friendly techniques. Online sample pretreatment responds to this trend, providing the development of effective approaches with low or no solvent and chemical consumption for the analysis of trace contaminants, such as pesticides. A TurboFlow method, for example, was successfully applied to the analysis of a wide spectrum of trace level pesticides, in drinking and surface water samples, taken from several sampling sites. Different TurboFlow columns were tested, showing that polymer-based columns offered the best performance [45, 46].

Pharmaceuticals are other relevant environmental trace contaminants. Their ecotoxicology is well known, while less is known about their metabolites and transformation products effects. Thus, multi-residues analytical methods are crucial to assess the risk of their presence in the environment. Furthermore, due to their low concentration, a pre-concentration procedure is often mandatory. Lopez-Serna and coworkers presented the development of an efficacious method for 58 pharmaceuticals and 19 metabolites/transformation products with an online pre-treatment based on turbo-flow chromatography in environmental aqueous samples. The combination of three TurboFlow columns in sequence showed the highest sensitivity [47].

6. Conclusion

Turbulent flow chromatography is a useful technique, able to remove potential interferences and to reduce preparation steps in an efficient way. In addition, the multiplex system allows to obtain faster results, by switching between different methods with minimum manpower, enhancing laboratory high-throughput productivity. Even if potentially heavy-matrix samples, as biological ones, could be directly injected into the TurboFlow column without any pre-purification, when performing quantitative analysis, in which the use of an internal standard is advisable, a simple addition of an organic solution of the internal standard to the samples could be performed, partially precipitating the protein content of the solution to be injected. In this way, the lifetime of the columns would be maximized. The resistance of the columns, granting hundreds of injections, and the possibility to integrate the TurboFlow to a pre-existent mass spectrometer, whose sensitivity is improved, guarantees also cost-effectiveness. Nonetheless, a high hardware and method development complexity have to be considered. This approach has only lately been increasing its applications, probably due to the complexity of the analytical methods development and optimization. Clinic and toxicology, where mass spectrometry must be handled even by not specialized laboratories, are the fields in which TurboFlow finds major application. However, TFC started to find use in food and environmental quality control, and it is intended to expand in these fields in the future.

Acknowledgements

The authors would like to thank Dr. Mariana Barcenas Rodriguez (European Molecular Biology Laboratory, Germany), Pr. Serena Indelicato (University of Palermo, Italy), and Pr. David Bongiorno (University of Palermo, Italy) for their advices and suggestions during the writing of this chapter and Dr. Claudio De Nardi (Thermo Fisher Scientific, U.S.), the most skilled TurboFlow user we know, for introducing us to this powerful system.

Author details

Francesca Di Gaudio[1], Annamaria Cucina[2*] and Sergio Indelicato[2]

1 PROMISE-Promotion of Health, Maternal-Childhood, Internal and Specialized Medicine of Excellence "G. D'Alessandro", Palermo, Italy

2 Chromatography and Mass Spectrometry Section, Quality Control and Chemical Risk (CQRC), Ospedali Riuniti Villa Sofia – Cervello, Palermo, Italy

*Address all correspondence to: annamariacucina1@gmail.com

References

[1] Dams R, Huestis MA, Lambert WE, Murphy CM. Matrix effect in bio-analysis of illicit drugs with LC-MS/MS: Influence of ionization type, sample preparation, and biofluid. Journal of the American Society for Mass Spectrometry. 2003;**14**(11):1290-1294. DOI: 10.1016/S1044-0305(03)00574-9

[2] Souverain S, Rudaz S, Veuthey JL. Matrix effect in LC-ESI-MS and LC-APCI-MS with off-line and on-line extraction procedures. Journal of Chromatography A. 2004;**1058**(1):61-66. DOI: 10.1016/j.chroma.2004.08.118

[3] Rouan MC, Buffet C, Marfil F, Humbert H, Maurer G. Plasma deproteinization by precipitation and filtration in the 96-well format. Journal of Pharmaceutical and Biomedical Analysis. 2001;**25**(5):995-1000. DOI: 10.1016/S0731-7085(01)00349-1

[4] Biddlecombe RA, Pleasance S. Automated protein precipitation by filtration in the 96-well format. Journal of Chromatography B: Biomedical Sciences and Applications. 1999;**734**(2): 257-265. DOI: 10.1016/S0378-4347(99)00355-2

[5] Kole PL, Venkatesh G, Kotecha J, Sheshala R. Recent advances in sample preparation techniques for effective bioanalytical methods. Biomedical Chromatography. 2011;**25**(1–2):199-217. DOI: 10.1002/bmc.1560

[6] Li W, Jian W, Fu Y. Basic sample preparation techniques in LC-MS bioanalysis. In: Sample Preparation in LC-MS Bioanalysis. Hoboken, New Jersey: John Wiley & Sons; 2019. pp. 1-30. DOI: 10.1002/9781119274315.ch1

[7] Picó Y, Fernández M, Ruiz MJ, Font G. Current trends in solid-phase-based extraction techniques for the determination of pesticides in food and environment. Journal of Biochemical and Biophysical Methods. 2007;**70**(2): 117-131. DOI: 10.1016/j.jbbm.2006.10.010

[8] Rogeberg M, Malerod H, Roberg-Larsen H, Aass C, Wilson SR. On-line solid phase extraction–liquid chromatography, with emphasis on modern bioanalysis and miniaturized systems. Journal of Pharmaceutical and Biomedical Analysis. 2014;**87**:120-129. DOI: 10.1016/j.jpba.2013.05.006

[9] Stoll DR, Carr PW. Two-dimensional liquid chromatography: A state of the art tutorial. Analytical Chemistry. 2017; **89**(1):519 531. DOI: 10.1021/acs.analchem.6b03506

[10] Dugo P, Cacciola F, Kumm T, Dugo G, Mondello L. Comprehensive multidimensional liquid chromatography: Theory and applications. Journal of Chromatography A. 2008;**1184**(1): 353-368. DOI: 10.1016/j.chroma.2007.06.074

[11] Ren Q, Wu C, Zhang J. Use of on-line stop-flow heart-cutting two-dimensional high performance liquid chromatography for simultaneous determination of 12 major constituents in tartary buckwheat (Fagopyrum tataricum Gaertn). Journal of Chromatography A. 2013;**1304**:257-262. DOI: 10.1016/j.chroma.2013.07.008

[12] Stoll DR, Shoykhet K, Petersson P, Buckenmaier S. Active solvent modulation: A valve-based approach to improve separation compatibility in two-dimensional liquid chromatography. Analytical Chemistry. 2017;**89**(17): 9260-9267. DOI: 10.1021/acs.analchem.7b02046

[13] Pirok BWJ, Stoll DR, Schoenmakers PJ. Recent developments in two-dimensional liquid chromatography: Fundamental improvements for practical applications. Analytical Chemistry. 2019;**91**(1): 240-263. DOI: 10.1021/acs. analchem.8b04841

[14] Couchman L. Turbulent flow chromatography in bioanalysis: A review. Biomedical Chromatography. 2012;**26**(8):892-905. DOI: 10.1002/ bmc.2769

[15] van Deemter JJ, Zuiderweg FJ, Klinkenberg A. Longitudinal diffusion and resistance to mass transfer as causes of nonideality in chromatography. Chemical Engineering Science. 1995; **50**(24):3869-3882. DOI: 10.1016/ 0009-2509(96)81813-6

[16] Victor P, Smuts TW. Turbulent flow chromatography. A new approach to faster analysis. Analytical Chemistry. 1966;**38**(2):274-281. DOI: 10.1021/ ac60234a030

[17] Rothfus RR, Prengle RS. Laminar-turbulent transition in smooth tubes. Industrial and Engineering Chemistry. 1952;**44**(7):1683-1688. DOI: 10.1021/ ie50511a055

[18] Edge T. Turbulent flow chromatography in bioanalysis. In: Wilson ID, editor. Handbook of Analytical Separations. Vol. 4. Oxford: Elsevier Science; 2003. pp. 91-112

[19] Quinn HM. High performance liquid chromatography method and apparatus. Published online July 6, 1999. Available from: https://patents.google.com/patent/ US5919368A/en. [Accessed January 9, 2023]

[20] Transcend and Aria systems. TurboFlow Method Development Guide. CH-106710 Revision 3 July 2009. Document provided by Thermo Fisher Scientific Inc. to its customers

[21] Michopoulos F, Edge AM, Theodoridis G, Wilson ID. Application of turbulent flow chromatography to the metabonomic analysis of human plasma: Comparison with protein precipitation. Journal of Separation Science. 2010; **33**(10):1472-1479. DOI: 10.1002/ jssc.200900789

[22] Chassaing C, Stafford H, Luckwell J, Wright A, Edgington A. A Parallel Micro Turbulent Flow Chromatography-Tandem Mass Spectrometry Method for the Analysis of a Pharmaceutical Compound in Plasma. Chroma. 2005; **62**(1):17-24. DOI: 10.1365/s10337-005-0562-3

[23] Gous T, Couchman L, Patel JP, Paradzai C, Arya R, Flanagan RJ. Measurement of the direct Oral anticoagulants Apixaban, dabigatran, edoxaban, and rivaroxaban in human plasma using turbulent flow liquid chromatography with high-resolution mass spectrometry. Therapeutic Drug Monitoring. 2014;**36**(5):597. DOI: 10.1097/FTD.0000000000000059

[24] Lindner JM, Vogeser M, Sorg K, Grimm SH. A semi-automated, isotope-dilution high-resolution mass spectrometry assay for therapeutic drug monitoring of antidepressants. Clinical Mass Spectrometry. 2019;**14**: 89-98. DOI: 10.1016/j.clinms.2019. 05.004

[25] Peake RWA, Hartigan CR, Esposito CL, et al. Multicenter evaluation of the thermo scientific prelude for measurement of immunosuppressant drugs using sample preparation liquid chromatography–Tandem Mass Spectrometry. Therapeutic Drug Monitoring. 2015;

37(2):161. DOI: 10.1097/FTD.0000000000000120

[26] Couchman L, Buckner SL, Morgan PE, Ceesay MM, Pagliuca A, Flanagan RJ. An automated method for the simultaneous measurement of azole antifungal drugs in human plasma or serum using turbulent flow liquid chromatography-tandem mass spectrometry. Analytical and Bioanalytical Chemistry. 2012;**404**(2): 513-523. DOI: 10.1007/s00216-012-6176-3

[27] Herviou P, Richard D, Roche L, et al. Determination of irinotecan and SN38 in human plasma by TurboFlow™ liquid chromatography–tandem mass spectrometry. Journal of Pharmaceutical and Biomedical Analysis. 2016;**118**: 284-291. DOI: 10.1016/j.jpba.2015.10.044

[28] Helfer AG, Michely JA, Weber AA, Meyer MR, Maurer HH. Liquid chromatography-high resolution-tandem mass spectrometry using Orbitrap technology for comprehensive screening to detect drugs and their metabolites in blood plasma. Analytica Chimica Acta. 2017;**965**:83-95. DOI: 10.1016/j.aca.2017.03.002

[29] Kasper DC, Herman J, De Jesus VR, Mechtler TP, Metz TF, Shushan B. The application of multiplexed, multi-dimensional ultra-high-performance liquid chromatography/tandem mass spectrometry to the high-throughput screening of lysosomal storage disorders in newborn dried bloodspots. Rapid Communications in Mass Spectrometry. 2010;**24**(7):986-994. DOI: 10.1002/rcm.4496

[30] Frederiksen H, Aksglaede L, Sorensen K, et al. Bisphenol A and other phenols in urine from Danish children and adolescents analyzed by isotope diluted TurboFlow-LC–MS/MS. International Journal of Hygiene and Environmental Health. 2013;**216**(6): 710-720. DOI: 10.1016/j.ijheh.2013.01.007

[31] López-García M, Romero-González R, Lacasaña M, Garrido FA. Semiautomated determination of neonicotinoids and characteristic metabolite in urine samples using TurboFlow™ coupled to ultra high performance liquid chromatography coupled to Orbitrap analyzer. Journal of Pharmaceutical and Biomedical Analysis. 2017;**146**:378-386. DOI: 10.1016/j.jpba.2017.08.026

[32] López-García M, Romero-González R, Garrido FA. Monitoring of organophosphate and pyrethroid metabolites in human urine samples by an automated method (TurboFlow™) coupled to ultra-high performance liquid chromatography-Orbitrap mass spectrometry. Journal of Pharmaceutical and Biomedical Analysis. 2019;**173**:31-39. DOI: 10.1016/j.jpba.2019.05.018

[33] Helfer AG, Meyer MR, Michely JA, Maurer HH. Direct analysis of the mushroom poisons α- and β-amanitin in human urine using a novel on-line turbulent flow chromatography mode coupled to liquid chromatography–high resolution-mass spectrometry/mass spectrometry. Journal of Chromatography A. 2014;**1325**:92-98. DOI: 10.1016/j.chroma.2013.11.054

[34] Schaefer N, Peters B, Schmidt P, Ewald AH. Development and validation of two LC-MS/MS methods for the detection and quantification of amphetamines, designer amphetamines, benzoylecgonine, benzodiazepines, opiates, and opioids in urine using turbulent flow chromatography. Analytical and Bioanalytical Chemistry.

2013;**405**(1):247-258. DOI: 10.1007/s00216-012-6458-9

[35] Ates E, Mittendorf K, Stroka J, Senyuva H. Determination of fusarium mycotoxins in wheat, maize and animal feed using on-line clean-up with high resolution mass spectrometry. Food Additives & Contaminants: Part A. 2013; **30**(1):156-165. DOI: 10.1080/19440049.2012.729162

[36] Ates E, Godula M, Stroka J, Senyuva H. Screening of plant and fungal metabolites in wheat, maize and animal feed using automated on-line clean-up coupled to high resolution mass spectrometry. Food Chemistry. 2014; **142**:276-284. DOI: 10.1016/j.foodchem.2013.07.054

[37] Roach JAG, DiBussolo JM, Krynitsky A, Noonan GO. Evaluation and single laboratory validation of an on-line turbulent flow extraction tandem mass spectrometry method for melamine in infant formula. Journal of Chromatography A. 2011;**1218**(28): 4284-4290. DOI: 10.1016/j.chroma.2011.05.025

[38] Mottier P, Hammel YA, Gremaud E, Guy PA. Quantitative high-throughput analysis of 16 (Fluoro)quinolones in honey using automated extraction by turbulent flow chromatography coupled to liquid chromatography—tandem mass spectrometry. Journal of Agricultural and Food Chemistry. 2008;**56**(1):35-43. DOI: 10.1021/jf072934d

[39] López-Gutiérrez N, del Aguilera-Luiz M, Romero-González R, JLM V, Garrido Frenich A. Fast analysis of polyphenols in royal jelly products using automated TurboFlow™-liquid chromatography–Orbitrap high resolution mass spectrometry. Journal of Chromatography B. 2014;**973**:17-28. DOI: 10.1016/j.jchromb.2014.09.038

[40] Presta MA, Bruyneel B, Zanella R, Kool J, Krabbe JG, Lingeman H. Determination of flavonoids and resveratrol in wine by turbulent-flow chromatography-LC-MS. Chromatographia. 2009;**69**(2):167-173. DOI: 10.1365/s10337-009-1132-x

[41] Fan S, Ma J, Cao M, et al. Simultaneous determination of 15 pesticide residues in Chinese cabbage and cucumber by liquid chromatography-tandem mass spectrometry utilizing online turbulent flow chromatography. Food Science and Human Wellness. 2021;**10**(1):78-86. DOI: 10.1016/j.fshw.2020.06.003

[42] Hollosi L, Mittendorf K, Senyuva HZ. Coupled turbulent flow chromatography: LC–MS/MS method for the analysis of pesticide residues in grapes, baby food and wheat flour matrices. Chromatographia. 2012;**75**(23): 1377-1393. DOI: 10.1007/s10337-012-2329-y

[43] Bousova K, Senyuva H, Mittendorf K. Quantitative multi-residue method for determination antibiotics in chicken meat using turbulent flow chromatography coupled to liquid chromatography–tandem mass spectrometry. Journal of Chromatography A. 2013;**1274**:19-27. DOI: 10.1016/j.chroma.2012.11.067

[44] Nathanail AV, Sarikaya E, Jestoi M, Godula M, Peltonen K. Determination of deoxynivalenol and deoxynivalenol-3-glucoside in wheat and barley using liquid chromatography coupled to mass spectrometry: On-line clean-up versus conventional sample preparation techniques. Journal of Chromatography A. 2014;**1374**:31-39. DOI: 10.1016/j.chroma.2014.11.046

[45] Asperger A, Efer J, Koal T, Engewald W. Trace determination of

priority pesticides in water by means of high-speed on-line solid-phase extraction–liquid chromatography–tandem mass spectrometry using turbulent-flow chromatography columns for enrichment and a short monolithic column for fast liquid chromatographic separation. Journal of Chromatography A. 2002;**960**(1): 109-119. DOI: 10.1016/S0021-9673(01) 01392-9

[46] Koal T, Asperger A, Efer J, Engewald W. Simultaneous determination of a wide spectrum of pesticides in water by means of fast on-line SPE-HPLC-MS-MS—A novel approach. Chromatographia. 2003;**57**(1): S93-S101. DOI: 10.1007/BF02492090

[47] López-Serna R, Petrović M, Barceló D. Direct analysis of pharmaceuticals, their metabolites and transformation products in environmental waters using on-line TurboFlow™ chromatography–liquid chromatography–tandem mass spectrometry. Journal of Chromatography A. 2012;**1252**:115-129. DOI: 10.1016/j.chroma.2012.06.078

[48] Mueller DM, Duretz B, Espourteille FA, Rentsch KM. Development of a fully automated toxicological LC-MSn screening system in urine using online extraction with turbulent flow chromatography. Analytical and Bioanalytical Chemistry. 2011;**400**(1):89-100. DOI: 10.1007/s00216-010-4560-4

Chapter 3

Monoliths Media: Stationary Phases and Nanoparticles

Mohamed Hefnawy, Ali El Gamal and Manal El-Gendy

Abstract

Monoliths media are gaining interest as excellent substitutes to conventional particle-packed columns. Monolithic columns show higher permeability and lower flow resistance than conventional liquid chromatography columns, providing high-throughput performance, resolution and separation in short run times. Monolithic columns with smaller inner diameter and specific selectivity to peptides or enantiomers have been played important role in hyphenated system. Monolithic stationary phases possess great efficiency, resolution, selectivity and sensitivity in the separation of complex biological samples, such as the complex mixtures of peptides for proteome analysis. The separation of complicated biological samples using columns is being revolutionized by new technologies for creating monolithic stationary phases. These techniques using porous monoliths offer several advantages, including miniaturization and on-line coupling with analytical instruments. Moreover, monoliths are the best support media for imprinting template-specific sites, resulting in the so-called molecularly-imprinted monoliths, which have an extremely high selectivity. In this chapter, the origin of the concept, the differences between their characteristics and those of traditional packings, their advantages and drawbacks, theory of separations, the methods for the monoliths preparation of different forms, nanoparticle monoliths and metal-organic framework are discussed. Two application areas of monolithic metal-organic framework and nanoparticle monoliths are provided.

Keywords: monoliths media, advantages of monoliths, hybrid monoliths, nanoparticle monoliths, monolithic metal–organic framework

1. Introduction

Liquid chromatography is a separation technique that uses a fluid phase through a porous bed. In order for the sample components to quickly equilibrate between the fluid and solid phases and spend a sizable portion of their entire residence durations in both, the fluid and bed materials must be selected carefully. The core of chromatography is a porous bed because the percolation of the mobile through the stationary phase is a crucial aspect of the chromatographic procedure. Solute molecule exchanges between the mobile phase stream and the stationary phase should occur quickly and often. Fast mass transfer kinetics, diffusion, close spacing within the bed, and a sizable surface area of contact between the stationary and mobile phases are all desirable characteristics for the column bed. High column efficiency is required if adsorption serves as the retention mechanism [1].

However, the bed's hydraulic resistance to the stream of the mobile phase should be kept at a reasonable level. The maximum length of the column that can be used with a specific pumping system will be impacted by a high hydraulic resistance. Fine particles should enhance separations in the early stages [2]. Unfortunately, there is not much that can be done to improve a packed bed's permeability. This permeability is determined by the packing density of the particles, which is best determined by the external porosity of the bed, and by their average size. With rising porosity, the permeability rises quickly in almost perfect proportion to its fifth power [3].

Now, imagine a porous bed that is the converse of a packed column bed. All the parts of the packed column bed result in a slower mobile phase flow velocity. The trend, so far, has been to prepare columns that have a high volume of through pores, hence a low hydraulic resistance leading to increase velocity. This makes it possible to prepare columns that have a high efficiency, because they can be long and yet can be operated with an acceptable head pressure, and are relatively fast, because they are operated at high flow velocity. In actual practice, a decrease in the column inlet pressure is far less attractive than a decrease in the elution time of the last eluted sample component or an increase in the column efficiency [4].

The popular English dictionary defines the monolith as a geological or technological feature such as a mountain or possibly a boulder, consisting of a single massive stone or rock [5]. Erosion usually exposes these formations, which are most often made of very hard and solid metamorphic rock. However, large pieces of rocks excavated from quarries, such as obelisks, are called artificial monoliths (**Figure 1**).

Figure 1.
Photograph of the porous monolith erected at the entrance of the Summer Palace Park, Beijing, China (Reprinted from Ref. [6]).

2. Monoliths

2.1 Definition

Monolith, in its broadest sense, refers to a column made out of a single substantial stone block. The name comes from the Greek words monos, which means "single," and lithos, which means "stone." It resembles a continuous single rod of porous material in chromatographic terms [7].

2.2 Advantages of monoliths

Due to the homogeneous distribution of macropores and mesopores across the network, which enables the separation of numerous analytes, monoliths are distinguished by a high permeability. While mesopores offer a significant surface area for separation and the permeability for solvent flow, macropores allow 100% of the mobile phase to pass through the column. In addition, packing like in a particle-packed column is not essential because the monolith can be created in place using poly-mediation. Due to the issue of the monolith shrinking in capillaries with greater i. d., this polymerization technique is only permitted in capillaries (typically less than 200 m in internal diameter i.d.). The distinctive characteristics of monolith, which have drawn the attention of numerous researchers and have been demonstrated to be an effective instrument in column technology [8], are high porosity and low back pressure. Pioneer efforts by Hjerten *et al.* [9] and Svec and Frechet [10] on developing polymer monolith and later by Minakuchi *et al.* [11] and Ishizuka *et al.* [12] developed silicon monolith have revolutionized the field of column technology.

Initially restricted to academic research, monolith is now acknowledged as a legitimate member of stationary phases. Other than permeability, monolithic columns are similar to conventional columns in terms of phase ratio and enantioselectivity [13]. The preparation and modification of monolithic columns to the correct porosity and pore diameter to meet various purposes are also simpler [14]. Through co-polymerization, specific selectors, such as chiral selectors, can be integrated into the monoliths and preserved there. In comparison to a particle column, the elution time can be cut by a factor of 5 to 10 [15]. All of these processes do not require any specialized knowledge, which makes interlaboratory investigations simple and comparable [13].

2.3 General characteristics of monoliths

Monolithic columns are made of a single piece of porous material that is hermetically sealed against the tube wall. As a result, the mobile phase stream is unable to bypass any appreciable portion of the bed and must instead percolate through it. A bimodal pore size distribution of the porous material is frequently used to describe it. This distribution's huge size mode corresponds to through pores or macropores. These pores are connecting into the channels that carry the majority, if not nearly all, of the mobile phase. In order to maintain the correct flow rate, the mobile phase must be pushed into the column at a head pressure determined by the average size of these channels, which also determines the column's permeability. Concerning the connection of the channels generated by these *via* pores, their constriction, and column permeability. The mesopore structure of the proton's polarons and all of these variables work together to determine the kinetics of mass transfer and column efficiency.

Practically no small size mode can be seen in the pore size distribution of the majority of polymeric materials [16–18].

Mesopores are represented by the small size mode of the pore size distribution of inorganic monoliths, such as silica. Between the channels that the through pores have created, there are lumps of porous material that contain these mesopores. For lack of a better term, these lumps are referred to as "porons." The lumps of porous material between the larger holes of particles with a bimodal internal porosity were first given this term [6]. The total porosity of the bed and its mechanical stability (the bed must withstand the pressure drop applied to cause the mobile phase to percolate through it) as well as the internal porosity and the precise surface area of contact between the two phases of the chromatographic system are correlated with one another. The latter regulates the solutes' retention factors and phase ratio. Therefore, the internal porosity of the column must strike a balance between the demands for small porons, a significant portion of the column's volume occupied by these porons, and a relatively large average size of the through pores [6]. More than two forms of pore size distribution may exist in monolithic materials. Because of their high adsorption energy and delayed mass transfer kinetics, micropores in any adsorbent have consistently had a negative impact on chromatographic performance. In polymer matrices and with silica-based adsorbents, microspores are uncommon [19], although they are prevalent in carbon-based adsorbents [20]. Another option is the mode associated with pores that have an average size that falls somewhere between the diameters of mesopores and through pores. This would be related to a certain class of monoliths with geometrical properties resembling those of the particles with bimodal porosity that were promoted for perfusion chromatography [21].

2.3.1 The through pores

Numerous research teams, including those of Nakanishi [22–24], Tanaka *et al.* [25], and later Cabrera [26, 27] and Leinweber [28], have explored the porous architectures of silica monoliths in depth. Phase separation is a byproduct of the acid-catalyzed hydrolysis and gelation of an alkoxysilane solution (such as tetramethoxysilane) in the presence of sodium polystyrene sulfonate, as demonstrated by Nakanishi and Soga [29]. Then, these researchers were able to create a solid mass of porous silica with a bimodal porosity by switching polyacrylic acid (PAA) with polystyrene sulfonate [22, 23]. Later, HPAA was effectively used to replace polyethylene oxide (PEO) [30]. Chromolith silica rod, a commercial item from Merck KGaA (Darmstadt, Germany), is depicted in **Figure 2** as having a porous structure [27]. By varying the solvent composition, the concentration of porogen, and the temperature, it is possible to modify the gel shape. The average size of the through holes grows with increasing porogen concentration (for example, HPAA). At low porogen concentrations, the big pores are not linked, whereas at high concentrations, distinct particles form [22]. Mercury intrusion porosimetry was used to estimate the average size of the through pores, which was found to be close to 1.7 m, and the external porosity or porosity of the through pores, which was calculated at 0.65 m [31].

2.3.2 The porons

The through pore network and the entire column of porons are filled with a continuous porous solid called porons. The monolith made using the method described by Nakanishi *et al.* [22, 24, 29] and bound to octadecyl dimethyl silyl groups

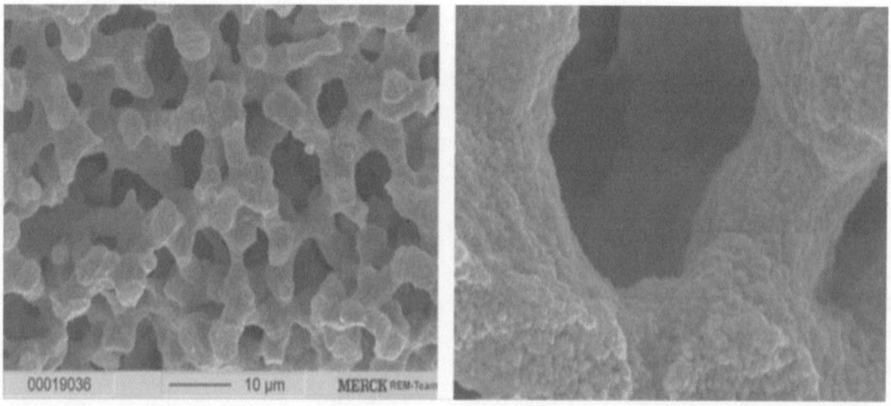

Figure 2.
Image of the porous structure of a typical monolithic silica column (left) and enlarged view of the entrance to a macropore or through pore (right) (Reprinted from Ref. [27]).

was shown by Nakanishi *et al.* [24] to be a porous solid with bimodal porosity. It was discovered that the porons had an average size of roughly 1 m. For neat silica, the mesopores or internal porosity are about 0.20 m [32].

3. Theory of separation

The most important characteristics of any chromatographic column are its permeability, which is determined by the average size of its through pores and external porosity, its efficiency, which is determined by the average sizes of its through pores and porons, the structure of the mesopore network, and the mobile phase velocity, and its capacity for retention (that is related to the specific surface area of the adsorbent, to the internal porosity and the pore size distribution). The monolithic beds prepared for liquid chromatography and the relationships between the characteristics of monolith beds and the permeability and efficiency of the prepared columns. Finally, the characterization of the monolithic columns and the conventional packed columns is used to compare these columns [33].

4. Separation mechanism

Hydrophilic interaction is hypothesized to be the main mechanism of hydrophilic interaction liquid chromatography (HILIC) separation for neutral polar molecules. The impact of the organic solvent concentration (mostly, acetonitrile, ACN) on the retention of small compounds (such as toluene, thiourea, and acrylamide) has frequently been studied in order to assess the HILIC capabilities of new stationary phases. The retention of polar molecules increases with increasing organic solvent content in HILIC stationary phases, which often exhibit typical HILIC behavior at high organic solvent content before returning to an apparently reversed phase (RP). When the stationary phases contain some hydrophobic characteristics and the organic solvent, content is below a threshold composition. In order to determine the polarity of HILIC stationary phases, one can use the critical composition of the mobile phase, which corresponds to the change from the HILIC to the RP mode [34]. The critical composition typically transitions to a lower concentration of ACN in the mobile phase with more polar stationary phases, permitting the use of mobile phases with higher water content in the HILIC mode. Porous zwitterionic monolithic columns exhibited very high selectivities for polar analytes, particularly tiny peptides, it was claimed [35–38]. Electrostatic repulsion-hydrophilic interaction chromatography was a new phrase proposed by Alpert [39] (ERLIC). Unique selectivity for charged polar analytes is provided by the interaction between hydrophilic contact and electrostatic repulsion. Numerous HILIC monolithic columns have shown similar ERLIC behavior. Ibrahim *et al.* [40] examined the impact of mobile phase salt concentration on the retention of two amino acids on coated silica monolith. Improving the quantity of methyl phosphonate decreases the electrostatic attraction between the cationic analyte and the cationic latexes, increasing the retention of the basic amino acid histidine. For the acidic amino acid, aspartic acid, however, raising the quantity of methyl phosphonate shields the electrostatic attraction, lowering its retention. The monolithic columns built with acrylamide also included an intriguing mixed-mode separation mechanism. A number of polar aromatic compounds were often used to analyze these relatively polar stationary phases [41]. The retention factors of the better-retained compounds were drastically reduced by a rise in mobile phase polarity, as would be expected for HILIC. However, some elution orders also changed, indicating that other mechanisms besides simple hydrophobic changes may be at play. This suggests that the elution order was not merely the inverted form of the reversed-phase separation [42]. It was reported that polar and charged analytes were retained by a mixed mode of mechanisms, including hydrophilic interaction, electrostatic interaction, H-bridging interaction, interaction, and possible aromatic adsorption [43, 44].

Similar to HPLC, the mechanism of mass transfer involves hydrophobic adsorption on the sorbent's surface and partition into bound alkyl chains. When transferring analytes to the adsorption process, physical parameters (such as surface area, average pore diameter, and pore volume of the adsorbents) are also taken into account. In general, the surface area of the adsorbent has a direct relationship with its adsorption capacity. The molecules adsorbed in the diffusion process have a relationship with the pore size. The saturation capacity of monolithic and packed silica columns was compared *via* frontal analysis for four distinct compounds, and the results showed that the monolithic columns had a 30–40% greater capacity while the binding constants were almost identical for both [45].

This shows that even while the chemistry of both surfaces is extremely similar, resulting in the same elution order for many analytes, the monolithic column's efficacy or accessible surface area is greater than that of the packed column. However, a conventional column of the same size has a far better load capacity since monolithic columns have a much lower density. In other words, within a certain velocity range, the shape of the breakthrough behavior of monoliths (BTCs) does not change. This indicates a very fast mass transfer and allows the speculation that the mass transfer is not dominated by the chromatographic velocity [45].

BTC shape is determined by equilibrium isotherm, mass transfer resistances, and axial dispersion [45].

Because the flow pattern can be represented as an axial dispersion plug flow, the differential fluid phase mass balance is:

$$-D_L\,\delta^2 c/\delta z^2 + u\,(\delta c/\delta z + \delta c/\delta t) + (1 - \acute{\varepsilon}/\acute{\varepsilon})\,\delta q/\delta t = 0 \tag{1}$$

with D_L the axial dispersion coefficient, q the sorbate concentration averaged over the particle, c the fluid phase concentration, z the column length, u the interstitial fluid velocity, t the time, and $\acute{\varepsilon}$ the voidage of the adsorbent bed.

The mass balance for an adsorbent particle yields the adsorption rate expression, which may be written as.

$$\delta q/\delta t = f\ (q, c) \tag{2}$$

The global mobile phase power model, apparent solid phase power model, or two-film model were all used to solve this equation. The mobility phase apparent total driving force model can be used to explain the mass balance of the adsorbed particle.

$$\delta q/\delta t = 6/d_p\ K_c\ (c\text{–}c^*) \tag{3}$$

or the apparent overall solid-phase driving force model.

$$\delta q/\delta t = 6/d_p\ K_q\ (q^*\text{–}q) \tag{4}$$

where c* is a fictitious concentration in the mobile phase that is in equilibrium with q, and q* is a fictitious concentration in the solid phase that is in equilibrium with c. The apparent total mass-transfer coefficients are Kc and Kq, respectively, while the surface-to-volume ratio for spherical particles is 6/dp. (c - c*) and (q* - q) are the apparent total driving forces, respectively. The overall apparent driving forces are, respectively, (c - c*) and (q* - q) [45].

The linear driving forces in the two-film theory are (c - c0) and (q0 - q), where subscript 0 designates a concentration at the two films' interface at which c0 and q0 are by definition in equilibrium, that is.

$$\delta q/\delta t = 6/d_p\ K_f\ (c\text{–}c_0) = 6/d_p\ K_s\ (q_0\text{–}q) \tag{5}$$

where k_f is a velocity-dependent external mass-transfer coefficient and k_s is the solid-phase mass-transfer coefficient. $1/k_f$ and $1/k_s$ are termed the external and solid-phase resistances.

Analytical formulas for the BTC can be established for various limiting forms of the isotherm, such as linear or rectangular, and axial dispersion is ignored.

BTCs can be predicted appropriately assuming that the equilibrium isotherm is rectangular when the actual equilibrium constant is greater than five [45].

At constant pattern, the mobile and solid-phase concentrations are related by Eq. (6).

$$q/q_F = c/c_F \tag{6}$$

where qF is the appropriate equilibrium concentration and cF is the feed concentration. The flow equations are transformed into ordinary differential equations under circumstances of the constant pattern. The constant pattern relation, Eq. (6), is changed to y = x because it is convenient to add the dimensionless concentrations x = c / cF and y = q / qF. The flow can be represented as a function of x by inserting Eq. (6) into Eqs. (3) and (4) and calculating c* and q*.

$$dq/dt = q_F\ dx/dt = 6/d_p\ K_c\ c_F\ \beta x(1\text{–}x)/1 + \beta(1\text{–}x) \tag{7}$$

and

$$dq/dt = q_F \, dx/dt = 6/d_p \, K_q \, q_F \, \beta x(1-x)/1 + \beta x \qquad (8)$$

where $\beta = bc_F$.

Inserting the Langmuir expression and the constant pattern relation in the two-film model, Eq. (5), yields.

$$\delta(x-x_0) = (1+\beta)x_0/1 + \beta x_0 - x \qquad (9)$$

where δ is the scaled mass-transfer resistance ratio.

$$\delta = k_f \, c_F / k_s \, q_F \qquad (10)$$

The interfacial mobile phase concentration without dimension x0 is obtained by solving Eq. (9), which in this instance may be done analytically [45]. The flux in the two-film model is expressed as a function of x by solving Eq. (9) for x0 and inserting it in Eq. (5). The solution is.

$$dq/dt = q_F \, dx/dt = 6/d_p \, k_f \, c_f \, (x-x_0) \qquad (11)$$

Despite having identical properties (i.e., particle sizes of less than 2 m), monolithic silica adsorbs the analyte with the narrow band on the column because its back pressure is significantly lower than that of the particle sorbent [46]. In contrast, the analyte adsorbed is quickly eluted by a little volume of elution solvent since the analyte is concentrated in the column. In order to provide adequate sample capacity and extraction efficiency, the standard solid-phase extraction cartridge often needs an elution volume that is two or three times the volume of the sorbent bed. Monolithic silica, on the other hand, only requires 100 L of eluent to elute the analyte adsorbed [46].

5. Types of monoliths

Based on the type of materials used in their preparation, monoliths are broadly categorized. Numerous monolith kinds may exist depending on this, although they are often divided into organic and inorganic-based monoliths. The chemistry of these two forms of monoliths serves as the basis for all other monolith types, whether through specific alterations or the use of many monomers. Organic monomers, like acrylamide [47, 48] and methacrylates [49, 50], are used for organic monoliths [51, 52], whereas inorganic monomers, like alkoxides of silicon [53, 54] are used for inorganic monoliths. The two categories differ in their chemistry of preparation, in which polymerization is applied for the organic monolith and hydrolytic polycondensation for the inorganic monolith.

5.1 Organic polymer monoliths

The idea of a monolith is thought to have originated from organic monoliths, which are composed of organic polymers. Developments in column technology have drawn a lot of interest since Hjerten *et al.* [9] introduced the first polymer monolith. His innovative work transformed the field of chromatographies by giving separating media new meaning and opening up new possibilities. They are a promising tool for separation in HPLC and capillary electrochromatography (CEC) modes due to the ease of fabrication, high permeability, porosity, and their ability to function in a wide pH range. They can be made using a variety of techniques, like radical polymerization (thermal or photoinduced) [55, 56], polycondensation reaction [57], and ring-opening metathesis polymerization [58, 59] depending on the type of stationary phase. Since the invention of the organic monolith, plenty of work has been carried out and it has been extensively reviewed. A mixture of monomers, a porogenic solvent, and an initiator are added to the chromatographic column or capillary to create polymer-based monoliths there. The final step is to treat the entire column or capillary to heat- or UV-initiated polymerization [60]. The initiator, which starts the polymerization of monomers, is converted into radicals. The leftover monomers and solvating solvents are distributed to the nuclei (precipitates) that the developing polymers create. The nuclei get bigger as the polymerization process goes on, which leads to clusters coalescing to produce a homogeneous integrated structure [61]. The monolith, which are uniform gels or stiff rods, is made up of interconnected clusters of globules [62]. The manufacture of organic polymer-based monolith typically involves the use of tubing with 8 mm i.d. capillaries for HPLC and 20–500 m i.d. capillaries for micro LC and CEC. Teledyne Isco, Inc. offers polymer monolithic columns for sale (IscoSwift®) [61]. Methacrylates, acrylamides, and styrene are frequently utilized as the monomers in organic monoliths [20]. Methacrylates monoliths have been thoroughly studied by Svec and his colleagues [55, 63]. Polystyrene was used to create a monolith by Wang et al. Several scientists then replicated their method to create the monolithic columns in a capillary style. The monolithic poly (polyphenyl acrylate-co-1,4-phenylene diacrylate) capillary columns were created by Bisjak *et al*. [64], and they investigated the impact of different polymerization parameters on the effectiveness of the separation. The monolith was created by thermally inducing the co-polymerization of phenyl acrylate and 1,4-phenylenediacrylate in the presence of azoisobutyronitrile, a radical initiator. The degree to which these parameters varied, including the choice and composition of the porogen, the temperature at which they polymerized, the amount of cross-linking monomer used, and the amount of initiator, affected how well the monolith separated. A reverse-phase HPLC system was used to test the produced monolith for the separation of proteins and oligonucleotides. In reverse-phase and ion-exchange chromatography, polymer-based monolithic columns are efficient for the high-speed separation of proteins, polypeptides, oligonucleotides, synthetic polymers, and some small molecules, but they exhibit relatively low efficiency for small solutes [51]. They can tolerate extreme changes in pH from 1 to 14 for reversing phase columns. Limitations include their relatively low mechanical stability and compared to conventional polymer beads, organic polymer monoliths provide higher efficiency and less swelling problems. The presence of micropores enhances analyte spreading which causes band broadening and finally leads to large mass transfer kinetics for small molecules [65].

5.2 Polar and nonpolar organic monoliths

Since they are known to have, among other things, the capacity to withstand a wide range of operational conditions and to offer solutions for a variety of separation problems, organic polymer monolithic stationary phases have attracted the attention of the separation science community in the past two decades. In order to adequately describe the numerous parameters involved in the manufacture of polar and nonpolar organic monoliths made of two distinct components, the author of this advanced review paper has attempted to do so. The first section discusses nonpolar organic monoliths, and the second section discusses polar organic monoliths and variations on them. The methacrylate/acrylate-based monoliths, styrene-divinylbenzene (DVB)-based monoliths, and hyper cross-linked monoliths are the three subgroups of the nonpolar organic monolith component. However, the section on polar organic monoliths, which primarily deals with monoliths derived from methacrylates, divides the monoliths into different categories based on the surface-charged groups they contain (for example, neutral, anionic, cationic, and zwitterionic monoliths) as well as how they were functionalized after polymerization [66].

5.2.1 Nonpolar organic monoliths

There are three commonly used nonpolar organic monolithic stationary phases, namely, acryl amide-based, acrylate/methacrylate-based, and styrene-based monoliths. The latter two are the most suited monoliths for reversed-phase chromatography (RPC) and RP-CEC applications.

5.2.1.1 Acrylate/methacrylate copolymer-based monoliths

Generally, the acrylate/methacrylate-based monoliths are more polar than their styrene DVB monolith counterparts. The formation of different kinds of monoliths is mainly by the choice of the functional monomers and cross-linking monomers. The structures of the functional and cross-linking monomers used for nonpolar and polar organic monoliths are shown in **Figure 3**.

5.2.1.2 Nature of functional monomers

The effect of the alkyl chain length and shape of the functional monomer on the structural features of the monolith has been reported. Rathnasekara *et al.* [67] used different alkyl methacrylate monomers with butyl, cyclohexyl,2-ethyl hexyl, lauryl, and stearyl functional groups in the polymerization mixture with ethylene glycol dimethacrylate (EDMA) as the cross-linker. In every instance, 1,4-butanediol (BDO) and 1-propanol were selected as the porogens. Test solutes included conventional proteins and alkylbenzenes, and typical metrics such as column permeability, methylene content, and phenyl selectivity were evaluated. The butyl methacrylate (BMA) monolith had a high level of permeability, whereas the lauryl methacrylate (LMA) monolith had a low level of permeability. Stearyl methacrylate (SMA)-based monolith had the best methylene selectivity, whereas BMA-based monolith displayed the highest phenyl selectivity. By using gradient elution in HPLC, lauryl and cyclohexyl methacrylate offered slightly improved separations for the tested standard proteins. Puangpila *et al.* [68] in CEC also looked at the impact of the functional monomers' alkyl chain length on solute retention. The method was first discussed by Okanda and El Rassi [69] and by Karenga and El Rassi [70]. The authors reported neutral monoliths (empty of fixed charges) to entirely prevent electrostatic interactions formed between charged solutes and the otherwise surface-attached charged moieties.

Using the identical cross-linking monomer pentaerythritol triacrylate, two distinct series of monolithic columns with surface-bound C8, C12, and C16 chains were created (PETA). While a series of monoliths (series B) was prepared by keeping the same composition of functional monomers and cross-linker, it produced chromatographic retention that increased as expected in the order of increasing the n-alkyl chain length. The monoliths (series A) were produced by adjusting the composition of functional

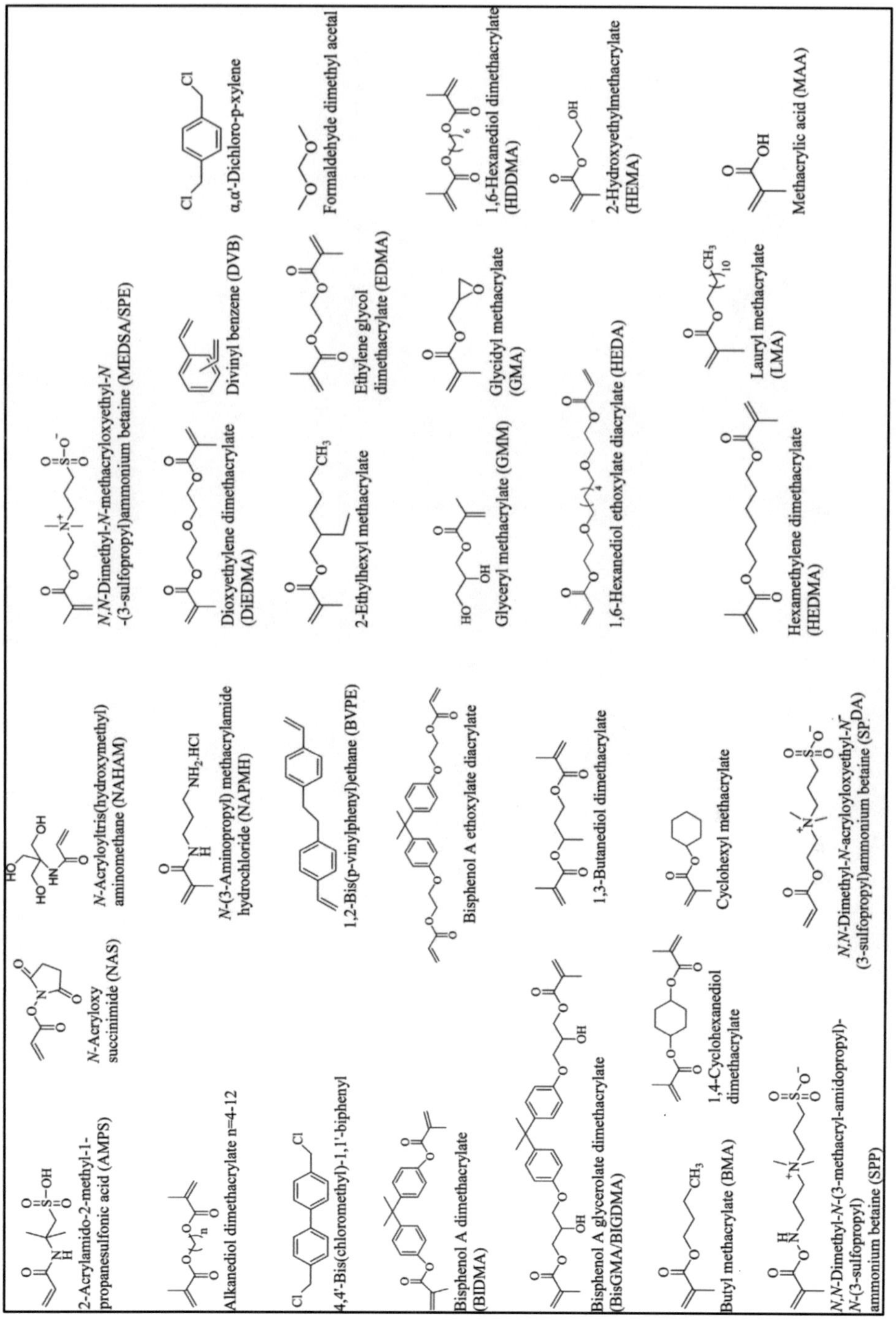
2-Acrylamido-2-methyl-1-propanesulfonic acid (AMPS)
N-Acryloxy succinimide (NAS)
N-Acryloyltris(hydroxymethyl) aminomethane (NAHAM)
N,N-Dimethyl-N-methacryloxyethyl-N-(3-sulfopropyl)ammonium betaine (MEDSA/SPE)
Alkanediol dimethacrylate n=4-12
N-(3-Aminopropyl) methacrylamide hydrochloride (NAPMH)
Dioxyethylene dimethacrylate (DiEDMA)
Divinyl benzene (DVB)
α,α'-Dichloro-p-xylene
4,4'-Bis(chloromethyl)-1,1'-biphenyl
1,2-Bis(p-vinylphenyl)ethane (BVPE)
2-Ethylhexyl methacrylate
Ethylene glycol dimethacrylate (EDMA)
Formaldehyde dimethyl acetal
Bisphenol A dimethacrylate (BIDMA)
Bisphenol A ethoxylate diacrylate
Glyceryl methacrylate (GMM)
Glycidyl methacrylate (GMA)
1,6-Hexanediol dimethacrylate (HDDMA)
Bisphenol A glycerolate dimethacrylate (BisGMA/BIGDMA)
1,4-Cyclohexanediol dimethacrylate
1,3-Butanediol dimethacrylate
1,6-Hexanediol ethoxylate diacrylate (HEDA)
2-Hydroxyethylmethacrylate (HEMA)
Butyl methacrylate (BMA)
Cyclohexyl methacrylate
Lauryl methacrylate (LMA)
N,N-Dimethyl-N-(3-methacryl-amidopropyl)-N-(3-sulfopropyl) ammonium betaine (SPP)
N,N-Dimethyl-N-acryloyloxyethyl-N-(3-sulfopropyl)ammonium betaine (SPDA)
Hexamethylene dimethacrylate (HEDMA)
Methacrylic acid (MAA)

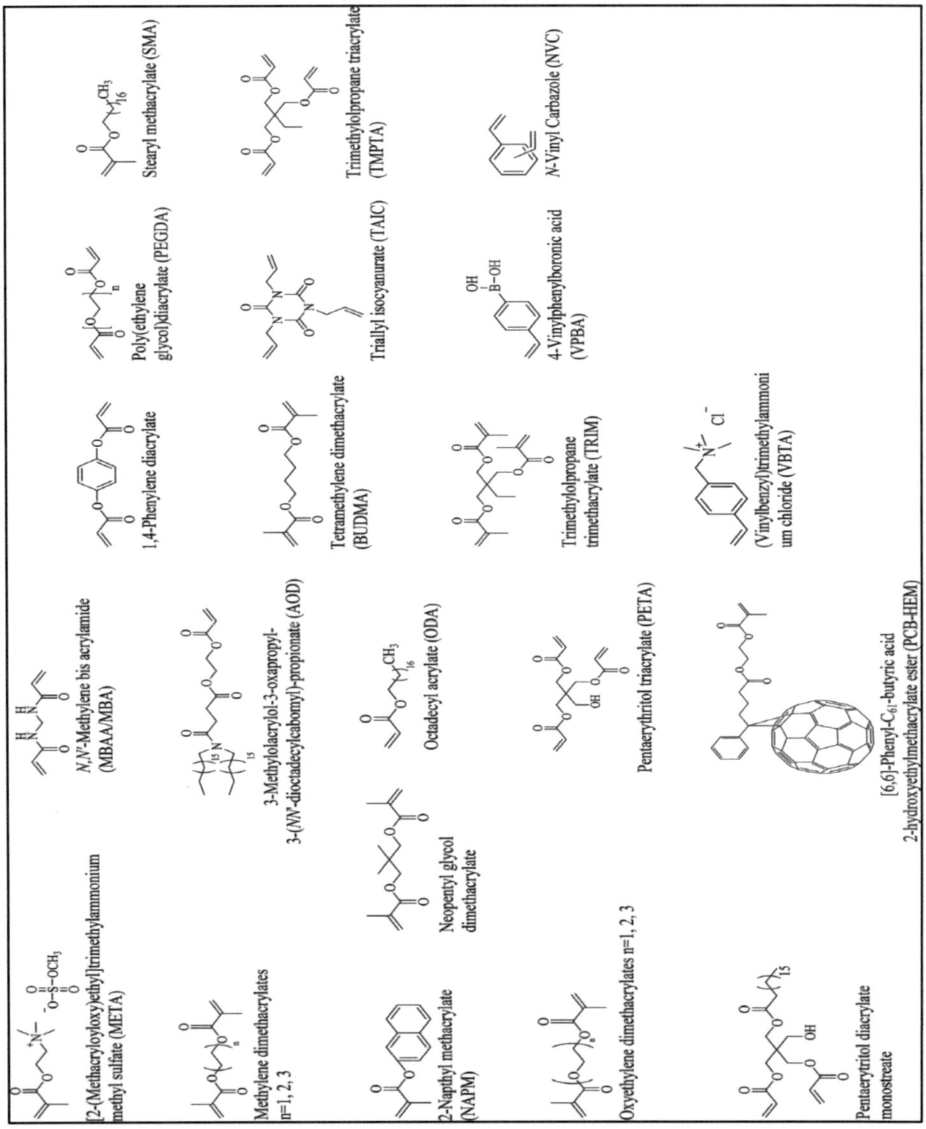

Figure 3.
Structure of functional monomers and crosslinked used in the preparation of nonpolar and polar monoliths, the monomers and cross-linked arranged alphabetically (Reprinted from Ref. [67]).

monomers and cross-linker to obtain comparable solute retention regardless of the alkyl chain length. The solvent was a ternary mixture of cyclohexanol, ethylene glycol, and water. The A series' C16 monolith produced the maximum separation performance for tiny solutes, however, the A column series was insufficient for separating proteins. The best separation efficiency for proteins was provided by the C8 monolith of the B series. The C16-monolith of the A series appears to offer the best separation for tryptic peptide mapping. An energetically "harder" C16 surface favored better separation of the smaller-size peptide solutes, whereas an energetically "softer" C8 surface permitted faster sorption kinetics and hence increased efficiency for big protein molecules. Briefly stated, proteins and peptides as well as neutral, polar, and charged solutes were effectively separated, and the outcomes were consistent with earlier research on neutral monoliths by Karenga and El Rassi [71].

Mixed ligand monolithic (MLM) columns were created for CEC as a revolutionary method of achieving unique selectivity [72]. The creation of these columns involved the copolymerization of various mixtures of the functional monomers octadecyl acrylate (ODA) and 2-naphthyl methacrylate (NAPM) in the presence of the cross-linker trimethylolpropane trimethacrylate (TRIM) and a ternary porogenic solvent made of cyclohexanol, ethylene glycol, and water. In the CEC of neutral, polar, and charged solutes, the combined retentive characteristic of the ODA ligand, which is merely hydrophobic, and that of the NAPM ligand, which is both hydrophobic and an interaction provider, was utilized. As anticipated, the MLM's makeup impacted how large the EOF was. In comparison to the NAPM monomer, the ODA ligand generally showed a higher affinity for the mobile phase ions. This is because the NAPM is a monomer based on methacrylate, whereas the ODA is based on acrylate. **Figure 4** illustrates the discovery that columns formed by either ODA or NAPM alone did not match the unique selectivity for a given set of solutes produced by columns with a specific mix of both ligands.

By co-polymerizing, the functional monomer 3-methylacrylol-3-oxapropyl-3-(N, N-dioctadecylcarbomyl)-propionate (AOD) with the cross-linker EDMA, Duan et al. [73] presented a unique monolithic column with double C-18 chains for RPC. The monolithic column for HPLC was created using fused-silica capillary columns (100 m id). A poly (SMA-co-EDMA) C-18 monolith was also created for the comparison investigations in accordance with the author's earlier work [74]. We employed a binary porogenic combination of BDO and 2-methyl-1-propanol. The back pressure

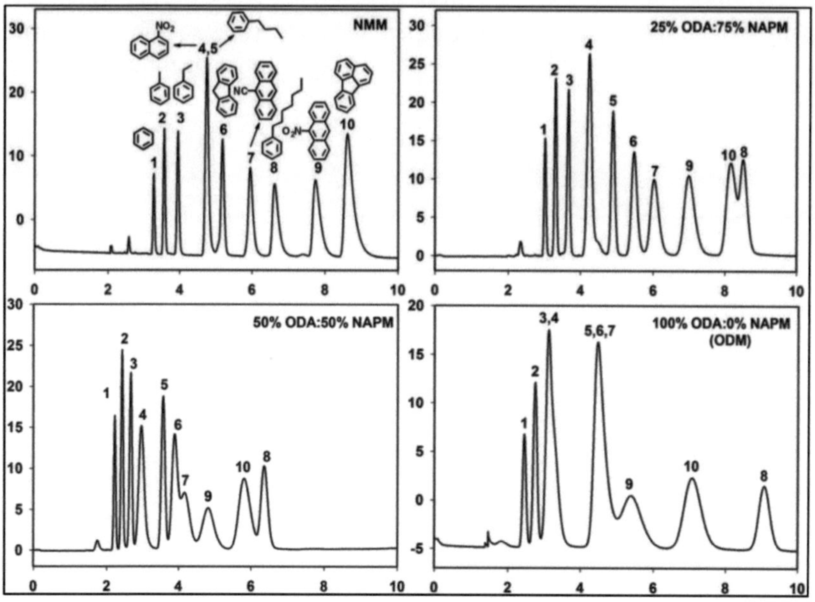

Figure 4.
Electrochromatograms showing the separation of five alkylbenzenes and five PAHs on monolithic columns with different mole fractions ODA/NAPM. Capillary column, 20 cm effective length, 27 cm total length × 100 μm id; mobile phase, 70% ACN, 1 mM sodium phosphate monobasic, pH 7.0, running voltage 20 kV; electrokinetic injection for 3 s at 10 kV. Solutes: 1, benzene; 2, toluene, 3,ethylbenzene; 4, 1-nitronaphthalene; 5, butylbenzene; 6, fluorene;7, 9-anthracenecarbonitrile; 8, heptylbenzene; 9, 9-nitroanthracene; and10, fluoranthene. EOF tracer, thiourea (Reprinted from Ref. [72]).

increased from 0.6 to 13.0 MPa at a flow rate of 500 nL/min as the monomer percentage changed from 50 to 70% w/w. Additionally, it was found that the back pressure increased from 0.6 to 13.0 MPa when the proportion of 2-methyl-1-propanol grew from 75 to 95% w/w. The results obtained were supported by the SEM images. The theoretical plate height of the poly (AOD-co-EDMA) monolith, which was 19.2 mm, was higher than that of the poly (SMA-co-EDMA) monolith, which was 32.1 mm at the same linear velocity of 0.85 mm/s. The permeabilities of this column with ACN, MeOH, and water were good. When strongly acidic and basic buffer were run over a sustained 30-hour period to test chemical stability, no noticeable deterioration was seen. The investigations into the batch-to-batch reproducibility studies and run-to-run repeatability research produced excellent results. In order to assess the effectiveness of the column and the behavior of the RP retention, respectively, Van Deemter plots and methylene selectivity tests were also performed. Using alkylphenols as test solutes, it was found that the methylene selectivity was 1.68, which was comparable to the 1.70obtained using the poly (SMAco-EDMA) monolithic column. Tocopherols (TOH) were utilized to test this column, and as can be shown in **Figure 5**, a complete separation of the, and TOH isomers was achieved in less than 30 minutes.

5.2.2 Polar organic monoliths

Despite offering a variety of chromatographic solutions for the separation, purification, and fractionation of a wide range of solutes, normal-phase chromatography (NPC), which employs polar stationary phases with nonpolar organic mobile phases, has its own limitations in the separation of highly polar analytes. Due to issues with hydrophilic materials dissolving in non-aqueous mobile phases, NPC of hydrophilic samples is challenging [75]. NPC has low repeatability for hydrophilic chemicals and poor mass spectrometry (MS) detection ionization efficiency [76]. Additionally, the

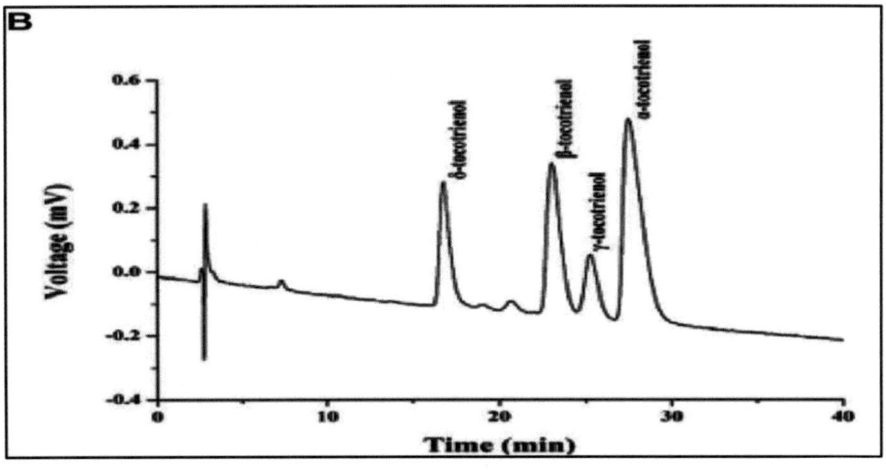

Figure 5.
A nano-LC separation of tocopherol homologs on the poly (AOD-co-EDMA) monolithic column. Conditions: Column dimensions 180 mm × 100 μm id; mobile phase: MeOH/H2O (93.5/6.5, v/v); detection, 292 nm; flow rate: 500 nL/min; injection volume: 20 nL (Reprinted from Ref. [73]).

majority of hydrophilic samples are poorly maintained on RP columns, eluting close to the column's dead time or displaying little to no resolution.

A chromatographic technique hydrophilic interaction liquid ion chromatography "HILIC" provides an alternative approach to effectively separate small polar compounds on polar stationary phases. In HILIC, polar analytes can be separated more effectively by combining a polar (hydrophilic) stationary phase with an organic-rich hydro-organic mobile phase. The inclusion of organic-rich mobile phase in HILIC offers some additional benefits that will increase the utilization of this method. These benefits include HILIC's appropriateness for direct coupling with MS detection and low column back pressure, which facilitates quick analyte separation with shorter analysis times [77]. The retention of polar solutes in HILIC and hydrophilic interaction CEC is strongly influenced by the makeup of the mobile phase (HI-CEC). HILIC/HI-CEC is typically operated with a low aqueous, highly organic mobile phase. The monolithic surface develops a water-rich layer under these HILIC circumstances. The adsorbed water layer on the surface of the stationary phase, the hydro-organic mobile phase, and the polar solutes in between are partitioned to create the separation. While HILIC frequently offers an efficient separation for polar analytes, a straightforward retention mechanism is not achievable for most molecules because hydrogen bonding, dipole–dipole, ion-dipole, and ion-ion interactions are also involved in addition to a partition mechanism [77].

During the past two decades, porous organic polymer monolithic columns have been widely employed as HILIC stationary phases due to their particular benefits, which include (i) high permeability, which reduces backpressure, (ii) low resistance to mass transfer, (iii) easy fabrication and surface functionalization, (iv) stability under severe pH values, and (v) a variety of functional monomers [78]. A variety of organic polymer monoliths with amino [79, 80], amide [81], hydroxyl [81], sulfoalkylbetaine [82], boronic acid [83], anilines and benzoic acids [84], and some other functionalities have been reported for use in HILIC/HI-CEC separations. **Table 1** summarizes the polar monoliths (neutral-cationic-anionic-zwitterionic monolith) and

Category	Name of monolith [poly (functional monoer-co-corss-linker)]	Porogens used	Column id/ material	Polymerization condition	Format/separated analyte
Neutral	Poly(GMM-co-PETA) (hydroxyl monolith)	Cyclohexanol, dodecanol, and water	100 μm id/fused silica	Thermal-initiation polymerization 60°C for 15 h with AIBN	HI-CEC/phenols, phenolic derivatives, nucleobases, and nucleosides
Neutral	Poly (NAHAM-co-PETA)	PEG in DMSO	100 μm id/fused silica	Thermal-initiation polymerization 60°C for 12 h with AIBN	CLC/ nucleosides, benzoic acid, anilines, nucleosides in urine
Cationic	Poly (NAPMH-co-EDMA) AP-monolith	Cyclohexanol, dodecanol, and methanol	100 μm id/fused silica	Thermal-initiation polymerization 50°C for 24 h with AIBN	HI-CEC/phenols, substituted phenols, and amides

Category	Name of monolith [poly (functional monoer-co-corss-linker)]	Porogens used	Column id/ material	Polymerization condition	Format/separated analyte
Cationic	Poly (VBTA-co-BisGMA)	Dodecanol, and methanol	100 μm id/fused silica	Thermal-initiation polymerization 60°C for 10 h with AIBN	HI-pCEC/benzoic acid derivative, phenols, nucleosides, and nucleobases
Cationic	Poly (META-co-PETA)	Cyclohexanol and ethylene glycol	100 μm id/fused silica	Thermal-initiation polymerization 60°C for 20 h with AIBN	pCEC/carboxylic phytohormones
Anionic	Poly(MAA-co-EDMA)	PEG and DMSO or dodecanol and toluene	100 μm id/fused silica	Thermal-initiation polymerization 60°C for 12 h with AIBN	cHILIC/anilines, benzoic acid derivatives, nucleosides, and tryptic digest of BSA
Anionic	Poly(AMPS-co-PETA)	Methanol ethyl ether and water	100 μm id/fused silica	Thermal-initiation polymerization 60°C for 20 h with AIBN	HI-CEC/neutral polar amides, phenol, and charge peptides
Anionic	Poly(GMA-co-VPBA-co-EDMA)	DEG and BDO	100 μm id/fused silica	Thermal-initiation polymerization 60 °C for 24 h with AIBN	CLC/alkaloids and protein
Anionic	Poly (HEMA-CO-MBAA)	DMSO, DMF, and water with Na_2HPO_4	100 μm id/fused silica	Thermal-initiation polymerization 60 °C for 24 h with AIBN	HI-CEC/nucleosides and some narcotic
Zwitterionic	Poly(SPP-CO-EDMA)	Methanol	100 μm id/fused silica	Thermal-initiation polymerization 70 °C for various time with AIBN	HILIC/benzoic acid derivatives, phenols, separation of ascorbic acid from dehydroascorbic acid
Zwitterionic	Poly(SPDA-co-MBA)/ poly(SPDA-co-EDMA)	Methanol and water	100 μm id/fused silica	Thermal-initiation polymerization 60 °C for 12 h AIBN	HILIC/ phenols, benzoic acid derivatives, peptides. Separation of alianation and urea in cosmetic products
Zwitterionic	Poly(SPE-co-PEGOA)	Isopropanol and decanol	75 μm id/fused silica	3 min under UV radiation with 2,2 dimethoxy-2-phenylacetophenone	HILIC/ amides benzoic acid derivatives and phenols
Zwitterionic	Poly(SPE-co-BVPE)	Methanol and toluene	200 μm id/fused silica	Thermal-initiation polymerization	HILIC gradient separation

Table 1.
Summary of polar monolith (neutral-cationic-anionic-zwitterionic monolith) and their applications for the separation of analytes.

their applications for the separation of analytes. Applications using polar organic monolith are shown in **Figures 6** and 7.

5.3 Monolithic metal: organic framework

5.3.1 A sol: Gel monolithic metal: Organic framework

The creation of a porous monolithic metal–organic framework (MOF) with a 259 cm^3 (STP) capacity following effective packing and densification. This is the greatest value for conforming shape porous solids documented to date, and it represents a more than 50% improvement above any experimental value previously reported. A significant advancement has been made in the use of mechanically robust conformed and densified MOFs for high volumetric energy storage and other industrial applications when nanoindentation tests on the monolithic MOF revealed robust mechanical properties, with hardness at least 130% greater than that previously measured in its conventional MOF counterparts [85].

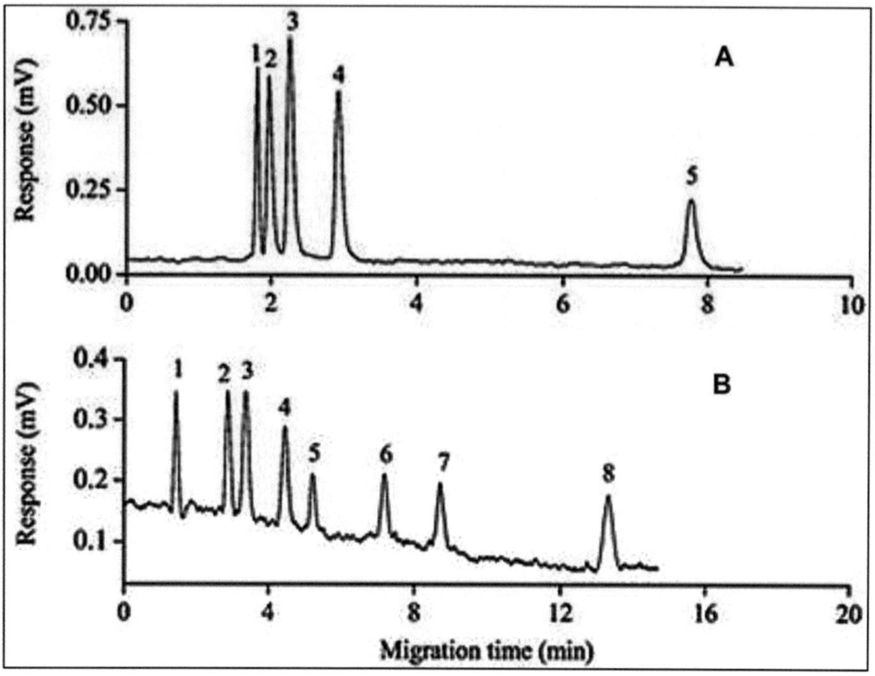

Figure 6.
Electrochromatographic profiling of phenols (A) and nucleic acid bases and nucleosides (B) on poly(META-co-PETA monolith. The profile (A) was recorded in the mobile phase 5 mM ammonium formate, pH 3.0, at 95% v/v ACN; pump flow: 0.02 mL/min; backpressure 250 psi; applied voltage: +10 kV. Solutes: (1) phenol; (2) catechol; (3) hydroquinone; (4) resorcinol; (5) pyrogallol. The profile (B) was recorded in the mobile phase 5 mM ammonium formate, pH 5.0, at 90% v/v ACN; the other conditions were the same as for (A). Solutes: (1) uracil; (2) adenine; (3) adenosine; (4) cytosine; (5) uridine; (6)guanine; (7) cytidine; (8) guanosine (Reprinted from Ref. [79]).

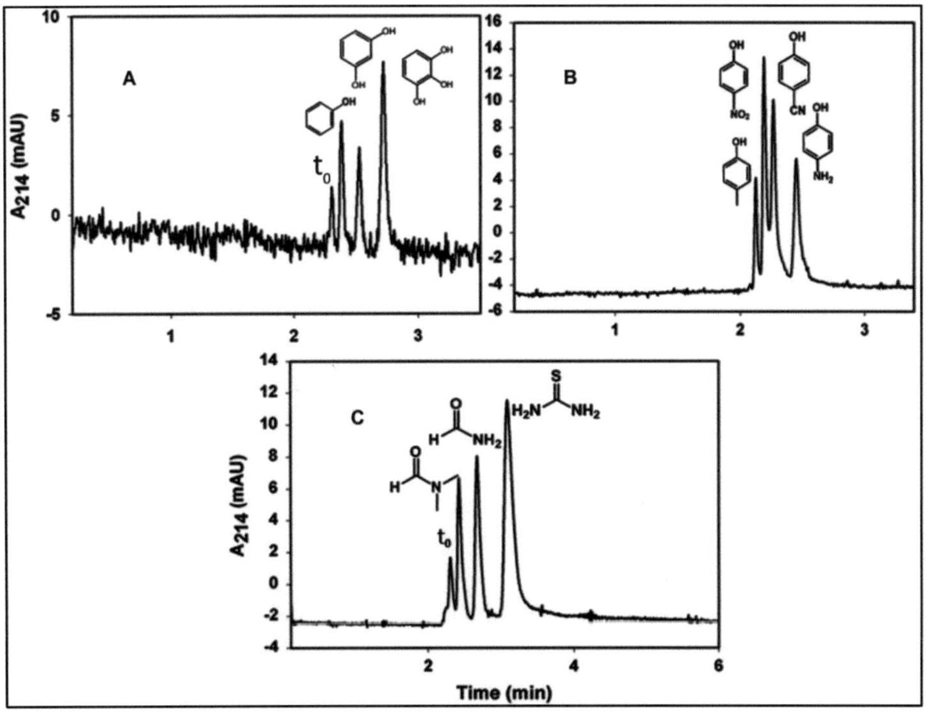

Figure 7.
Separation of three phenols (A), four phenol derivatives (B), and three other compounds (C) obtained on the neutral hydroxymonolith. Condition: Hydro-organic mobile phase, 5 mM NH4Ac (pH 8.0) at 95% ACN (v/v), running voltage 20 kV, column temperature 20°C, sample injection, pressure at 5 bar for 10 s. solutes in C are DMF, formamide, and thiourea (left to right) (Reprinted from Ref. [81]).

5.3.2 Metal–organic frameworks in biomedicine

5.3.2.1 Influence of the composition

For biomedical applications, porous materials must have a composition that is conducive to living things. There are not many toxicity studies that involve MOFs or coordination polymers at this time. Most of the information is limited to the assessment of the individual toxicity of the metals and linkers. Of course, only metals with a tolerable level of toxicity should be taken into account here. The choice to exclude a certain composition from biomedical use could be based on a number of factors, including the application, the trade-off between risk and benefit, and the kinetics of degradation, biodistribution, accumulation in tissues and organs, and excretion from the body. All metals and linkers could therefore be employed in these applications, although at varying doses based on the aforementioned requirements. According to their oral lethal dosage 50 (LD50), the most suitable metals at first look are Ca, Mg, Zn, Fe, Ti, or Zr, whose toxicity ranges from a few g/kg to more than 1 g/kg (calcium). The most popular one is the use of exogenous linkers, either made synthetically or naturally from substances that do not interfere with bodily cycles. Relevant exogenous MOFs for bio-applications include those made from iron (III) polycarboxylates, such as MIL-100(Fe) [86], zinc adeninate-4,40 biphenyldicarboxylate BioMOF-1 [87, 88], and magnesium coordination polymers, such as the magnesium 2,5-dihydroxoterephthalate CPO-27(Mg) (CPO for Coordination Polymer from Oslo) [89] (**Figure 8**).

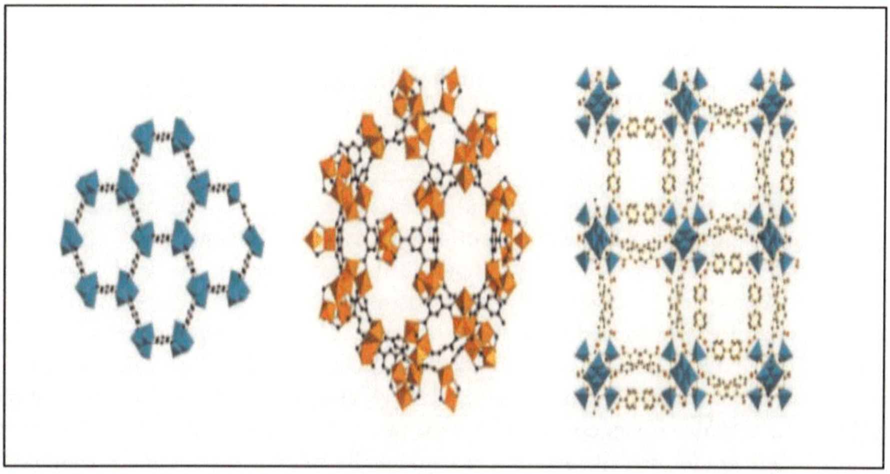

Figure 8.
View of the structures of a few topical MOFs, here CPO-27(Mg, Zn) (left), MIL-100 (Fe) (center), and Bio-MOF-11 (right), based on exogenous linkers for bioapplications. Metal polyhedra and carbon atoms are in blue (Zn, Mg) or orange (Fe), and black, respectively (Reprinted from Ref. [88]).

These solids have large pores (4–29) and surfaces between 1200 and 2200 m^2. Biomolecules (NO, CO, H_2S, medicines, etc.) may coordinate strongly at these accessible Lewis acid sites to better control the release [90–92]. The structures of a few MOFs based on endogenous linkers such as Bi citrate [93], Mg formate [94], Cu aspartate [95], Zn-dipeptide [96], Fe gallate [97], Fe Fumarate [98] and Fe muconate [99] are shown in **Figure 9**.

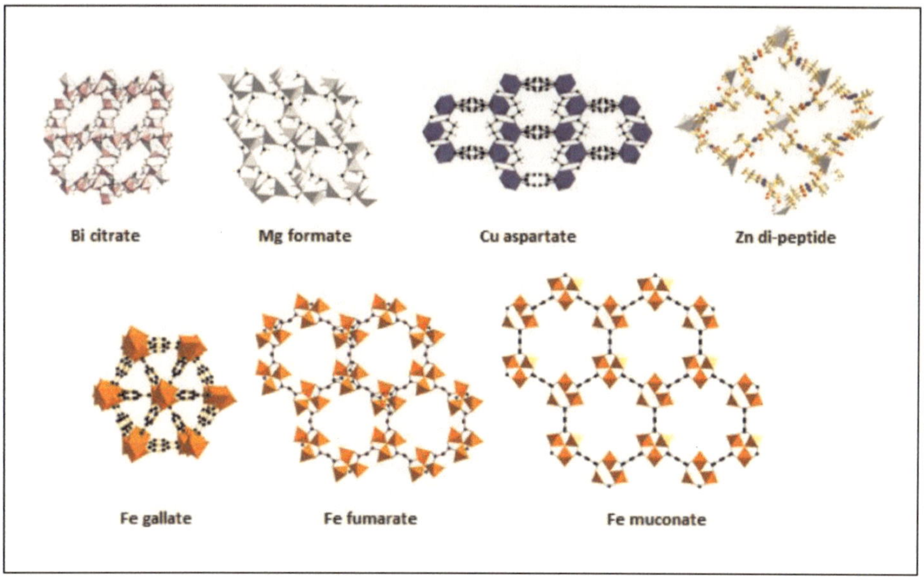

Figure 9.
View of the structures of a few MOFs based on endogenous linkers such as; Bi citrate [93], Mg formate [94], Cu aspartate [95], Zn-dipeptide [96], Fe gallate [97], Fe fumarate [98], and Fe muconate [99]. Metal polyhedral are in pink, gray, gray, blue, or orange (for Bi, Mg, Zn, Cu, and Fe, respectively) and carbon atoms in black, respectively.

Exogenous linkers should be eliminated from the body following the *in vivo* delivery of the MOFs in order to prevent potentially harmful side effects. As a result of their high polarity and ease of removal under physiological conditions, typical polycarboxylic or imidazolate linkers are not initially thought to be very toxic, with rat oral doses of 1.13, 5.5, and 8.4 g/kg for terephthalic, trimesic, 2,6-napthalenedicarboxylic acid, and 1-methylimidazole, respectively.

The use of functionalized linkers to adjust MOFs' absorption, distribution, metabolism, and excretion is yet another potential application for exogenous linker-based MOFs (ADME). Additionally, the presence of functional groups inside the framework can alter the host-guest interactions for the adsorption and distribution of therapeutic molecules, enabling a better control of the release. While a number of porous MOFs based on modified linkers that contain polar or apolar functional groups such as amino, nitro, chloro, bromo, carboxylate, methyl, and perfluoro are known, the most prevalent functionalized systems involve porous metal terephthalates based on iron or zinc [100]. One could also cite a series of organically modified porous Zn imidazolate solids [101]. In the case of functions, flexible MOFs will not only modify the host and guest interactions, but will also drastically affect the flexibility of the MOF during the adsorption or delivery of the biomolecule [102, 103] (**Figure 10**).

5.3.2.2 Nanoparticle monoliths

For some administration routes, where highly exact sizes are needed, particle size is a limiting constraint. For instance, the parenteral method necessitates stable

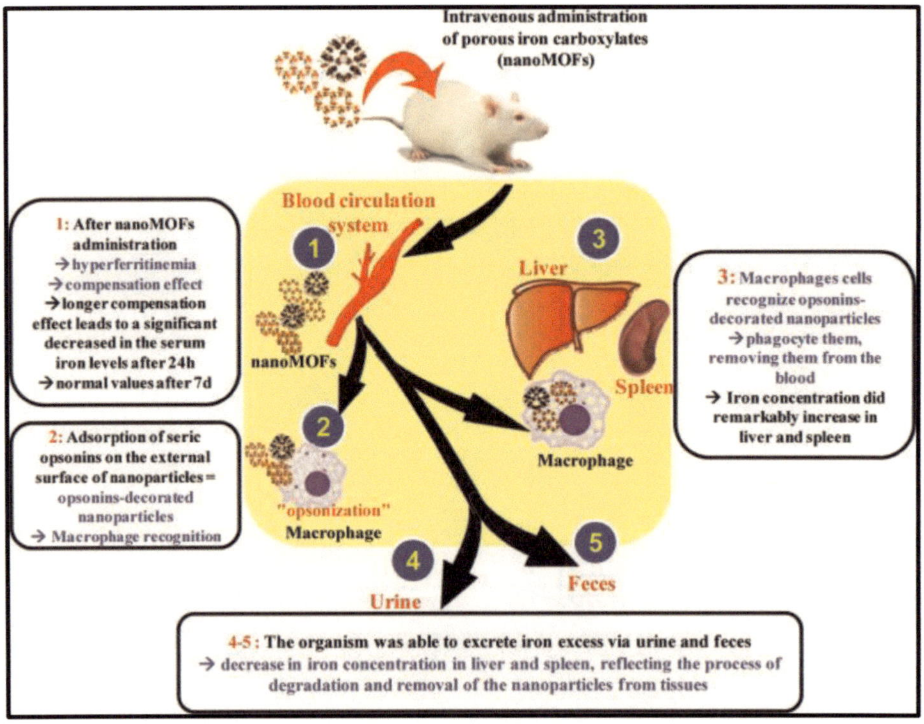

Figure 10.
Scheme of biodistribution of iron nanoMOFs according with the iron concentration (Reprinted from Ref. [88]).

solutions or suspensions of nanoparticles smaller than 200 nm in order for them to freely flow through the tiniest capillaries. Therefore, the creation of homogenous, monodispersed, and stable nanoparticles is a significant issue that has been addressed thus far using the following techniques:

5.3.2.3 The conventional hydro/solvothermal route

The conventional hydro/solvothermal method depends on a number of parameters, including reaction duration, temperature, stoichiometry, dilution pH, additives, etc. Topical examples include the porous zinc terephthalate MOF-5 (100–200 nm) [104] or the flexible porous Iron (III) dicarboxylates MIL-88A (150 nm) [105], MIL-88B-4CH3 (40 nm) [105], obtained by reducing reaction time or temperature, either employing low temperature or atmospheric pressure conditions. Alcohols can also be used to create nanoparticles of the zinc imidazolate ZIF-8 (40 nm) or the porous iron muconate MIL-89 (30 nm) [106] at low temperatures. Although particle sizes less than 100 nm are frequently achieved, the lack of homogenous and efficient heating typically results in a significant drop in yield and a high degree of polydispersity because the nucleation and growth stages are not under control [107]. Acidobasic or inhibitory additives (acetic acid, hydroxybenzoic acid, and pyridine) [106] are frequently used to adjust the reaction kinetics or slow the nucleation development process. Pyridine has been suggested as an inhibitor in the solvothermal production of porous Indium terephthalate particles by Cho *et al.* [108]. Acetate ions were utilized by Horcajada *et al.* [106] to create tiny nanoparticles of the malleable, porous iron muconate MIL-89 as growth inhibitors (30 nm). Similar results were obtained by Tsuruoka et al. [109] with porous Cu2-(naphthalene dicarboxylate) 2(1,4-diazabicyclo (2,2,2) octane nanorods. A polyvinylpyrrolidone (PVP) polymer was used by Kerbellec *et al.* [110] to produce extremely tiny nanoparticles of a luminous terbium terephthalate (about 4–5 nm) at ambient pressure and room temperature. This approach was additionally used for various lanthanide (Tb, La, Tm, or Y) terephthalate MOF luminous micro- and nanoparticles [111].

5.3.2.4 Reverse-phase microemulsions

Reverse-phase microemulsions are based on a metal source, an organic linker, and micelles of cationic cetyltrimethylammonium bromide surfactant (CTAB) in a nonporous Ln [112–114] or Mn [115] based polycarboxylates MOFs with interesting imaging properties. This technique allows a control of particle size by tuning the dimensions of the micelles. Isooctane/1-hexanol/water mixture led to nonmetric nonporous Ln [112–114] or Mn [115] based polycarboxylates MOFs with interesting imaging properties technique allows a control of particle size by tuning the dimensions of the micelles.

5.3.2.5 Sonochemical synthesis

Micro and nanoscale MOFs have recently been synthesized using the quick, simple, and ecologically friendly sonochemical synthesis technique. Acoustic cavitation is caused by ultrasonic irradiation and results in the bursting of bubbles, localized hot patches, a significant temperature/pressure differential, and quick molecular movement. This encourages the development of high-energy microreactors, which causes MOFs to crystallize quickly [116, 117]. While other porous crystalline structures have been successfully created at the nanometric scale under ultrasonic circumstances, the microporous flexible iron(III) terephthalate MIL-5376 and a stiff copper trimesate have been crystallized at the microscopic scale [118, 119]. The Seo research group has detailed the ultrasonic synthesis of porous copper trimesateHKUST-1 in DMF (10–200 nm) and zinc trimesate in ethanol (50–100 nm) at room temperature and ambient pressure using selective organoamine sensing [120]. The flexible porous iron fumarate MIL-88A's particle size can be varied between 100 and 740 nm, with the lowest particle sizes being obtained by using inhibitors, extremely high dilutions (0.01–0.008 M), or low temperatures (°C). As a result, only low yields (5 wt%) are produced [107].

5.3.2.6 Microwave-assisted hydro/solvothermal synthesis

For the creation of MOF nanoparticles, microwave-assisted hydro/solvothermal synthesis is an effective, homogeneous, and quick technique. The high thermal efficiency of polar solvents results in local superheated regions and high dielectric absorptivity, which favors a rapid and homogeneous nucleation process over crystal formation [121]. This technique was used to produce the zinc terephthalates IRMOF-1, 2, and 3 (100 nm) [122] and the mesoporous chromium terephthalateMIL-101 ($\sim$22 nm) [123, 124]. Sefcjk and McCormick [125] have extensively studied the flexible microporous iron terephthalate MIL-53 (350–1000 nm), the mesoporous iron trimesate MIL-100 (200 nm), the iron aminoterephthalate MIL-101NH2 (120 nm) and the iron fumarate MIL-88A ($\sim$20 nm), as shown in **Figure 8**.

5.4 Silica monoliths

Sol–gel synthesis, which enables extraordinary control over the composition and shape of the generated monolith, is typically used to create silicon alkoxide-based monoliths. This method is flexible and may be used to create a variety of monoliths. Starting with Si(OR)4, the reaction creates a siloxane (Si-O-Si) network in the polymer. Phase separation, condensation, and hydrolysis compete with one another to create a porous monolith [125]. There were several steps involved in creating the silicon alkoxide-based monolith. Alkoxy groups are replaced by hydroxyl groups in the first step of the process, which also results in the synthesis of silanol groups (SiOH) and alcohol. These extremely reactive silanol groups condense with additional alkoxy silanes or with one another in the next steps, creating a siloxane bond. Following these preparatory phases, a soil's precursors hydrolyze under acidic or basic conditions, forming a gel with a three-dimensional network following a series of condensation steps. To create the monolithic matrix, this gel ages and passes through phase separation under a certain set of circumstances [125]. Tetramethoxysilane (TMOS) and tetraethoxysilane (TEOS) are typically the two precursors that are employed the most in the preparation. By regulating the rate of condensation and hydrolysis, the reactivity of the precursors can regulate the distribution of subunits throughout the network. Several species are produced as a result of the reaction, and they all proceed through hydrolytic polycondensation at different speeds than the precursor. When compared to other precursors, TMOS undergoes rapid hydrolysis, making it one of the most reactive [126]. When compared to TEOS, it was found that TMOS produces pores that are more consistent, thinner, and have a higher surface area. Various strategies have been developed to address the issues of cracking and shrinkage. Despite various concessions in their performance or stationary phase qualities, they proved to be useful. The initial method that distinguished the development was the creation of particle-loaded monoliths in a capillary format with the goal of creating a column without cracks. This method of preparation was first presented by Dulay *et al.* [127] with the idea of embedding particles inside the holes or cavities made inside the produced matrix. Using sol–gel technique, monoliths were created by embedding ODS particles (3–5 m) in tetraethoxysilane (TEOS). A 40 cm long, 75 m capillary was filled with the solution, and the resulting monolith was employed in CEC. A combination of aromatic and non-aromatic chemicals was used to assess the column's performance. Due to the ODS particles' deep embedding and non-homogeneous packing, which shields the particles from the analyst, different efficiency is attributed to these factors. Bakry *et al.* [128] adapted the above technique and enclosed silica particles within the polymeric backbone to address these issues (polystyrene divinylbenzene). In their process, silica particles were first packed into prepared silica capillary using the slurry packing method, and then an immobilizing fluid made up of styrene, divinyl benzene, azobisisobutyronitrile (AIBN), and decanol was added. After polymerization, the created monolith underwent a test to determine whether polyphenols, peptides, and proteins could be separated. Sometimes it is simpler to synthesize monoliths made in capillaries than bigger diameter columns because they can form a covalent bond with the inner capillary wall, which gives the monolithic bed more stability.

5.5 Hybrid monoliths

The sol–gel method, which entails a succession of hydrolysis and polycondensation stages to generate a three-dimensional network structure, is also used to prepare organic–inorganic hybrid monolithic materials [129]. Except for the co-polymerization of a second precursor, which is in charge of adding an organic moiety to the silica backbone to prepare a desirable stationary phase, the preparation techniques are the same as in the single alkoxide production. The addition of one or more modified precursors, such as 3-mercaptopropyltrimethoxysilane (MPTS) [130], sets it apart from pure organic and inorganic-based monoliths. The organic moiety is uniformly distributed throughout the matrix and connected to the expanding silicon network *via* a non-hydrolyzable Si-C bond, which improves the performance and efficiency of the monolith to the desired chromatography. Allyl, octadecyl, phenyl, and amino groups were adsorbed onto the matrix, eliminating the need for the time-consuming and laborious post-matrix modification issue [131–133].

As a replacement for the current stationary phases, a hybrid monolith made utilizing a sol–gel process and hybrid materials have been proposed. This hybrid monolith has shown promise because it offers more benefits than a traditional silica monolith. In order to increase column efficiency, stability, and selectivity, sol–gel hybrid materials are specifically created to have desired qualities and eliminate the ones that are undesired. Additionally, hybrid monoliths can be created directly by combining organic and inorganic monomers for the desired interaction, negating the need for the functionalization of stationary phases, which is more typical with the conventional method, which involves creating the monolith first and then functionalizing it. Therefore, Hayes and Malik [134] created the monoliths using the solution without using particles and without the necessity for frits in their hunt for a substitute to combine the preparation and functionalization of silica monolith in a single step. They provided a straightforward method for creating functionalized porous monoliths that are chemically adhered to the silica capillary's inner walls. The sol–gel precursor,

N-octadecyldimethyl [3-(trimethoxysilyl) propyl] ammonium chloride, was employed to create porous monoliths helpful for CEC studies with chemically bound ODS ligands. Later, employing methyltrimethoxy silane (MTMS) alone as a precursor, Laschober *et al.* [135] developed a capillary monolith based on the sol–gel process, which has the advantage of having increased hydrolytic stability of the Si-C bond. Additionally, they looked at how different parameters affected the monolith's morphology, which was shown by the size of its pores, skeleton, and surface. The prepared monolith is usable throughout a wider pH range. Due to their decreased compatibility with polar solvents, the monoliths produced by MTMS showed a higher tendency for spinoidal decomposition than tetraalkoxysilanes. Furthermore, unlike the case with tetraalkoxysilanes, synthesis does not call for the functionalization of the capillary walls or a hydrophilic polymer like polyethylene glycol. At pH 1, the reaction is conducted to create monoliths with bicontinuous morphologies. When compared to tetralkoxysilane, the surface area is less, which may be due to the involvement of only three of silicon's four valences in forming bonds with other organo-silica tetrahedrons. The monolith's chromatographic performance was evaluated using each component separately, and separation was predicted based on the components' various retention times. Dunn and Zink [136] recently evaluated the characteristics and uses of molecules trapped in a silica matrix made using the sol–gel method. In order to explore the impacts of changes in many factors, such as solvent composition, polarity, viscosity, pH, and speeds of chemical reaction, entrapped molecules acting as a spectroscopic probe offer insight into sol–gel chemistry. As probes, a variety of organic compounds have been used to track the chemical changes in sol–gels.

6. Preparation of hydrophilic monolithic columns

6.1 Silica-based monolithic columns

The Nakanishi and Soga team invented silica-based monolithic columns in the early 1990s [137]. The majority of these silica-based monoliths that have been described were created using methods that are comparable and involve the hydrolysis of one or more silanes, primarily tetramethoxysilane, in an acidic solution while the presence of an appropriate porogen. A sufficient mesopore is formed when the gel is aged and macerated in a basic solution. The gel is then dried and heated before having its surface modified with the appropriate ligands. One particular benefit of the silica-based monolithic column is that its macropores and mesopores can be tailored separately to achieve the best performance for specific analytical objectives. This is in addition to the two main generic advantages of all monolithic columns, namely, the high permeability and the low resistance to mass transfer. According to Jiang *et al.* [138], the majority of articles on silica monolithic columns dealt with applications based on commercial columns. He hypothesized that this was due to the fact that several stages (such as drying and cladding of the rods) are challenging to overcome in academic laboratories and that Merck, the producer, has carefully crafted patents designed to close any gaps [139]. The Chromolith Performance monolith, one of Merck's commercially available silica monolithic columns, has the potential for HILIC applications because it features a polar silica surface. However, based on these Chromolith Performance materials, just one such HILIC application has been documented. This column was employed by Pack and Risley [140] for the detection and quantification of lithium, sodium, and potassium in the HILIC separation mode.

Additionally, certain monolithic capillary columns made entirely of silica have been created in academic labs and used for HILIC. The first silica-based monolithic capillary columns were created at the end of the 1990s [137] using the sol–gel method described by Nakanishi and Soga. However, the drying and calcination processes involved in its preparation take time. Additionally, Puy *et al.* [141], a 2 h water washing phase, which may remove 70% of the polyethylene glycol, was used in place of the 24 h calcination step (PEG). The monolith's stability, morphology, or chromatographic qualities are unaffected by the residual PEG. Later, they discovered that monolithic capillaries made of pure silica and employed in HILIC mode or normal phase mode did not require the hydrothermal treatment step at 120°C. The hydrothermal treatment was suppressed, which increased the retention factors but did not affect CEC or nano-LC efficiency [142]. The monolithic silica matrix synthesized from a sol–gel process was chemically modified by 3-aminopropyltrimethoxysilane, 3-(2-aminoethylamino) propyltrimethoxysilane [138], or diethylenetriaminopropyltrimethoxy silanc [143] in order to produce a column for hydrophilic interaction applications. The surface modifications were simply carried out by a solution of

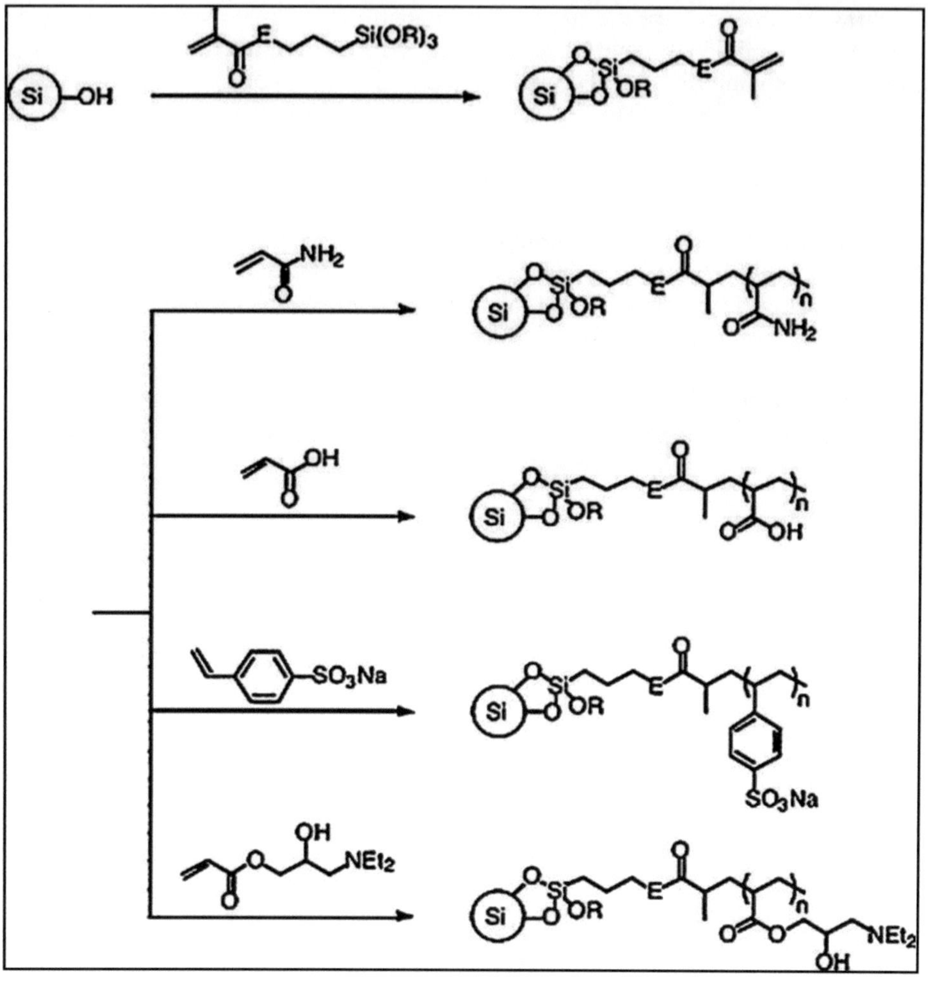

Figure 11.
Synthetic scheme of the polymer-coated monolithic silica columns (Reprinted from Ref. [144]).

pumping the silane in anhydrous toluene through the bare silica monolithic capillary column for 1 h which was thermostated at 75°C or 110°C for various times. These amino silica monolithic stationary phases exhibited HILIC behavior toward neutral solutes. However, no comparative research between these columns and bare silica monoliths has been reported. Moreover, Ikegami *et al.* [144] summarized a polymer coating procedure in their review article as shown in **Figure 11**.

6.1.1 Functionalization of silica-based monoliths

The study on the modification or functionalization of silica-based monolithic columns has been developed to satisfy the varied needs for the separation of complex substances in LC and CEC based on the silica-based monolithic columns created utilizing the aforementioned procedures. **Figure 12** illustrates how monolithic columns can entirely independently manage the chemical and porous properties by post-modification. Additionally, post-modification allows for the avoidance of the need to prepare a monolithic column's porous characteristics from scratch each time a change in chemical functionality is sought. For the functionalization of porous silica-based monoliths, a number of approaches have been identified. The far simpler method to change the silica-based monolithic columns, as shown in **Figure 13** [127], was to use various silane chemicals.

7. Selected monoliths media applications

Applications developed for monoliths media have been tremendous, and it would be impossible to review them all in this chapter book. We have, however, categorize studies of monoliths media with metal–organic framework and nanoparticle monoliths in chromatography separations. Some representative examples published over the last decade are summarized below to provide the reader with an appreciation of the possible breadth of applications.

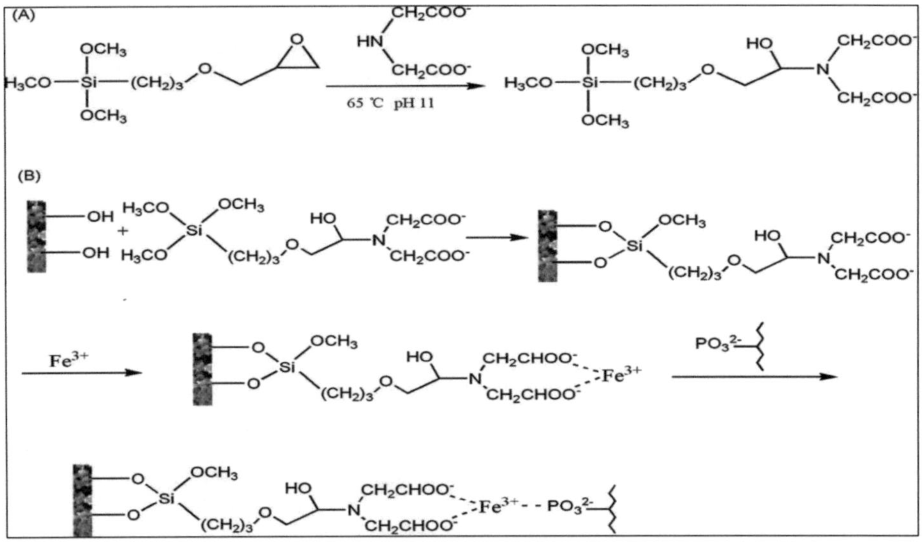

Figure 12.
Post-modification of silica monolith (Reprinted from Ref. [144]).

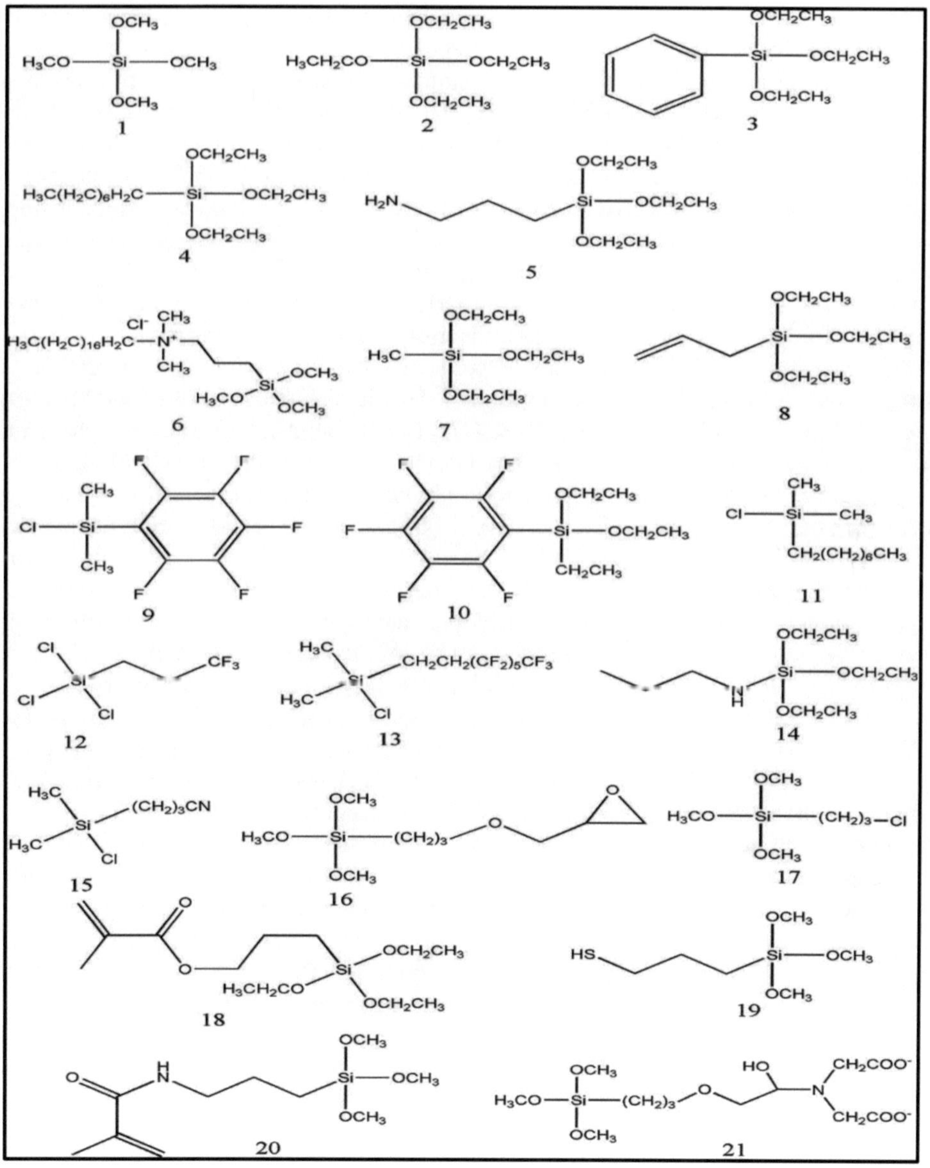

Figure 13.
List of silane reagents for silica-based monolithic stationary phases (Reprinted from Ref. [127]).

7.1 Applications of MOFs in LC

The use of MOFs for the LC separation of compounds, such as various aromatic molecules [145–147], medications [148], dyes [149, 150], pollutants [151], and peptides [152], has attracted a lot of interest. In fact, the early MOF-based HPLC separation of xylene isomers and ethylbenzene employed MIL-47 as the stationary phase. It was then employed to extract styrene and ethylbenzene [153]. Other MOFs that have been proposed as HPLC stationary phases include ZIF-8 [154], MIL-53 [155, 156], MIL-101 [157], MIL-47 [158], and MIL-53 [155, 156]. Direct packing of MOF particles, on the other hand, led to low separation efficiency, undesired peak morphologies, and large column backpressures due to their irregular forms, sub-micrometer size, and wide size ranges. It can be challenging to regulate the pore size of silica deposition substrates and the synthesis of MOFs [159]. One method utilized to separate xylene isomers, dichlorobenzene, and chlorotoluene isomers, as well as ethylbenzene and styrene, was a MIL-101(Cr) packed column (5 cm 4.6 mm i.d.) [157]. For ethylbenzene, the packed column's efficiency was just 20,000 plates per meter, which is not particularly noteworthy. In comparison to the meta- and para-isomers, the affinities for o-dichlorobenzene and o-chlorotoluene were greater. These early MOF publications frequently omitted the fact that alternative, frequently commercial, LC columns had been able to separate these compounds more effectively for years.

The deposition of MOF thin films onto core-shell silica particles was regulated using a layer-by-layer deposition technique [160]. In this work, MOFs such MIL-101 (Fe) NH2 (Fe3(O)(BDC NH2)3(OH)(H2O)2, BDC = benzene-1,4-dicarboxylic acid, and UiO-67 (Zr6O4(OH)4(BPDC)6, BPDC = 4,4′ -biphenyldicarboxylate) were employed. Mixed-mode liquid chromatography was used to manufacture and analyze metal–organic framework-assisted hydrogel-bonded silica composite microspheres (Zn-BTC MOF, DNs-hydrogel@SiO2) [161]. A more tolerable efficiency for this kind of chromatographic column is 90,300 plates per meter (sucrose). Alkylbenzenes, organic acids, carbohydrates, a few antibiotics, polycyclic aromatic hydrocarbons, insecticides, and anions could all be isolated from nucleosides and their bases (**Figure 14**). All of them have, of course, been readily separable on a variety of commercial stationary phases for many years. A chiral sorbent made of a Zn + 2-based MOF with L-lactic acid ligands was utilized to extract a variety of sulfoxide enantiomers [162]. Additionally, it was utilized in chiral alkyl aryl sulfoxides' traditional LC separation. In open tubular capillary electrochromatography, a different D-(+)-camphoric acid + Zn + 2-based MOF was utilized to separate the enantiomers of flavanone and praziquantel [163]. Additionally, several achiral compounds were isolated. Magnetic particles were utilized in conjunction with another chiral MOF to

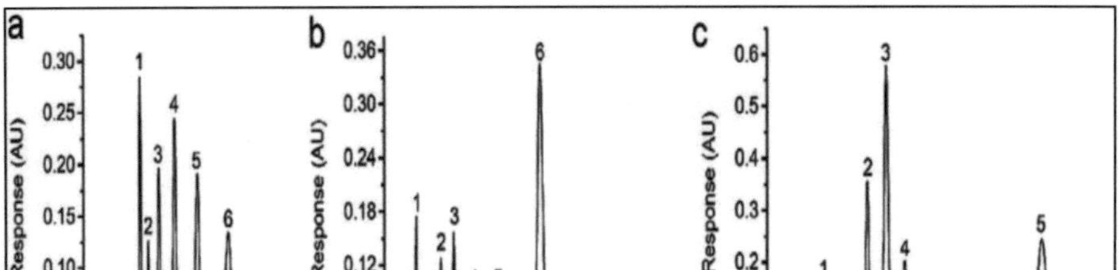

Figure 14.
Chromatograms for the separation of (a) alkylbenzenes, (b) polycyclic aromatic hydrocarbons, and (c) pesticides on DNs-hydrogel@SiO$_2$ column. Analytes: (a) 1. Benzene, 2. Methylbenzene, 3. Ethylbenzene, 4. Propylbenzene. 5, butylbenzene. 6, Pentylbenzene; (b) 1. Benzene, 2. Naphthalene, 3.2-methylnaphthalene, 4, fluorene, 5. Phenanthrene, 6. Fluoranthene, 7. Benzoanthracene; (c) 1. Flufenoxuron, 2. Meturon, 3. Chlortoluron, 4. Diuron, 5. Diflubenzuron. Mobile phase: (a) ACN/H$_2$O (40/60, v/v); (b) ACN/H$_2$O (38/62, v/v) and (c) ACN/H$_2$O (35/65, v/v). UV detection at 254 nm. Column temperature: 25°C. flow rate: 1.0 mLmin-1 (Reprinted from Ref. [161]).

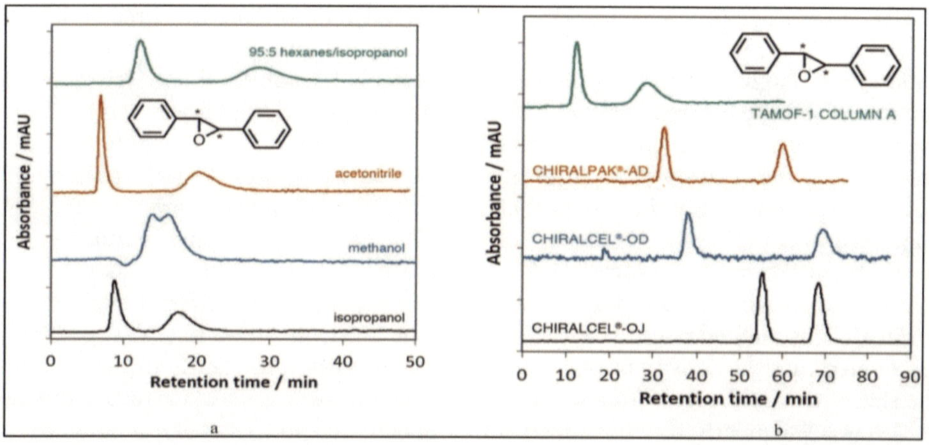

Figure 15.
Representative chromatograms for the separation of trans*-2,3-diphenyloxirane enantiomers with HPLC column packed with TAMOF-1: (a) separation with different mobile phases using column A: Isopropanol, methanol, acetonitrile, and 95:5 hexanes/isopropanol (v/v). (b) Comparison with commercial columns CHIRALPAK®-AD, CHIRALCEL®-OD and CHIRALCEL®-OJ in 95:5 hexanes/isopropanol (v/v) (Reprinted from Ref. [165]).*

enantioselectively enrich an enantiomeric pharmacological intermediate from the solution [164].

It has been claimed that the porous and durable homochiral (MOF) TAMOF-1, which was made from copper (II) and a natural L-histidine linker, can separate a few racemic combinations, including several medications [165]. There have been claims that it performs better than some commercial HPLC chiral columns, but the outcomes do not appear to back up even this extremely superficial comparison (**Figure 15**). For the enantioseparation of 18 racemates, including alcohols, phenols, amines, ketones, and organic acids, a similar homochiral D-his-ZIF-8@SiO2 composite was tested [166]. The separations were carried out using n-hexane/isopropanol as the solvent in the standard phase mode. For reversed phase separations in capillary electrochromatography, a monolithic column made of 1-allyl-methylimidazolium chloride (AlMeIm + Cl-) copolymerized with ZIF-8 was utilized [167]. This method produced monolithic columns that were used to separate neutral chemicals, anilines, and phenols. The performance of the column efficiency was improved by the synergistic interaction between the ionic liquid and ZIF-8. The maximum column efficiency was found in toluene, which measured 2.07×10^5 theoretical plates m^{-1}.

7.2 Applications of nanoparticle-based monoliths

7.2.1 Gold nanoparticles

Due to their simple and inexpensive synthesis, huge surface area, molecular recognition properties, and compatibility with living things, gold nanoparticles (GNPs) are among the nanomaterials that have been the subject of the most research. GNPs have been used in a variety of contexts and are receiving more interest in the monolithic sphere. A new silica monolithic stationary phase functionalized with octadecanethiol GNPs for CEC was created by Ye *et al.* [168]. In response to neutral solutes, the resulting GNP-modified silica monolith displayed normal reversed-phase electrochromatographic activity. Due to the great affinity of gold for these moieties, GNPs can be covalently linked to surfaces containing amino, thiol, or cyano functionalities [169]. A-glucosidase was then easily and stably immobilized onto GNPs due to the strong affinity between gold and enzyme amino groups, which was achieved by the same group [170] by covalently attaching GNPs to the surface of a polymer monolith *via* the creation of an Au-S bond. Additionally, silica monoliths coated with bovine serum albumin-gold nanoparticles (BSA-GNPs) conjugate were reported to be employed as chiral stationary phases for the CEC enantioseparation of various phenylthiocarbamyl amino acids [171]. Lv and colleagues [172] created a novel method for creating porous polymer monoliths with improved GNP coverage of pore surfaces. This method considerably improved the immobilization of GNPs, and it was discovered that the density of pore surface covering was significantly influenced by the size of the GNPs. As a "universal" intermediate ligand, the surface of the attached gold was subsequently functionalized with 1-octanethiol and 1-octadecanethiol to produce the desired monoliths for the separation of proteins in reversed-phase mode. In **Figure 16**, three proteins are separated in a gradient of acetonitrile (ACN), with the order of elution determined by the hydrophobicity of the proteins. Monoliths modified with 15, 20, and 30 nm NPs provided the best separations because these sizes produced the densest covering of the pore surface with GNPs. The combination of GNPs attached by layered architecture to hypercrosslinked polymer-based monoliths and functionalized with hydrophilic properties was initially described by the same group [173]. With its use in the separation of tiny polar analytes, such as nucleosides and peptides, under hydrophilic interaction chromatographic conditions, this effective monolithic stationary phase was proven.

7.2.2 Silver nanoparticles

The advantages of silver nanoparticles (AgNPs) over other nanostructured metal particles, such as their advantageous electrical conductivities [174], antibacterial activities [175], and optical properties [176], are well established. Monolithic columns with embedded AgNPs were created and tested in order to combine the unique properties of AgNPs with monoliths [177]. Investigated how the presence of AgNPs affected the polymer matrix's morphological and chromatographic characteristics. AgNPs have a very strong affinity for the anions of heavy halogenides like astatide and iodide. In order to remove these constituents from solutions, AgNPs may, therefore, be a great candidate [178]. The creation of a macroporous monolithic column with anchored AgNPs for the removal of extra radioiodine from a radiolabeled pharmaceutical was demonstrated by Sedlacek et al. [179].

7.2.3 SiO_2/TiO_2 nanoparticles

As a result of their better chemical affinities and amphoteric characteristics, metal oxides have been introduced into monolithic structures with increasing effort. Although TiO_2 has received the greatest attention and is the most often utilized metal oxide, few research studies have successfully created monolithic capillary columns made entirely of pure TiO_2 [180, 181]. Without a doubt, creating monoliths based on TiO_2 presents a number of difficulties. Additionally, according to several related reports, SiO_2/TiO_2 composite materials outperform pure TiO_2 materials in the enrichment of phosphorylated targets [182–184]. Wang *et al.* [185] developed a facile sol–gel method for preparing a novel SiO_2/TiO_2 in-capillary composite monolith under mild

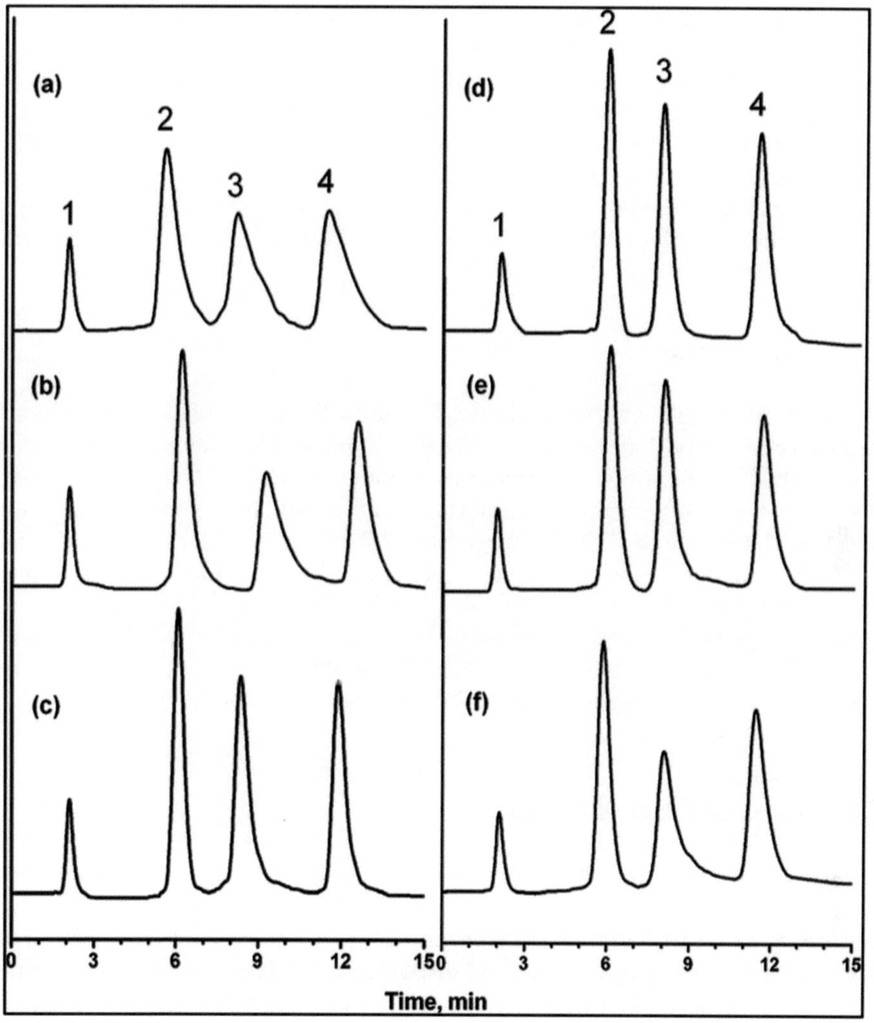

Figure 16.
Reversed-phase separation of proteins using monolithic columns containing GNPs modified with 1-octanethiol. Conditions: Columns: (a) 5 nm GNP, 100 mm × 100 mm I.D.; (b) 10 nm GNP, 110 mm _ 100 mm I.D.; (c) 15 nm GNP, 110 mm × 100 mm I.D.; (d) 20 nm GNP, 108 mm × 100 mm I.D.; (e) 30 nm GNP, 80 mm × 100 mm I.D.; and (f) 40 nm GNP, 89 mm × 100 mm I.D.; mobile phase: A 0.1% aqueous trifluoroacetic acid, B 0.1% trifluoroacetic acid in ACN, gradient from 20 to 70% B in A in 20 min, flow rate: 1.0 ml min^{-1}, injection volume: 10 nl, detection wavelength: 210 nm. Peaks: (1) impurity from ribonuclease A, (2) ribonuclease A, (3) cytochrome C, and (4) myoglobin (Reprinted from Ref. [172]).

conditions, which was successfully applied to specifically capture phosphopeptides as a metal–oxide affinity chromatography material.

7.2.4 Zirconia nanoparticles

Zirconia is a desirable and appropriate substitute for silica as the support, primarily due to its exceptional and one-of-a-kind chemical, mechanical, and thermal stabilities [186–188]. The benefits of zirconia NPs include the fact that they are stable at temperatures up to 200°C and exhibit no discernible evidence of disintegration over a wide pH range. Zirconia-based stationary phases can avoid certain notable limitations of silica-based stationary phases, such as restricted use in a constrained pH range and temperature range. Zirconia's distinct surface chemistry, meanwhile, expands the range of chromatographic separation applications [189]. A porous zirconia monolith (ZM) modified with cellulose 3,5-dimethylphenylcarbamate (CDMPC) was created by Kumar and Park [190] and employed as a chiral stationary phase for the CEC separation of a group of fundamental chiral chemicals.

Subsequently, they reported other zirconia-based chiral stationary phases for the CEC separation of β-blockers [191], chiral acids and bases [192], basic compounds [193], and other chiral analytes [194].

7.2.5 Silica nanoparticles

Liu *et al.* [195] developed a novel cage-like silica NP-functionalized silica hybrid monolith based on the "one-pot" method using thiolene click chemistry. The resulting hybrid monolithic column was characterized and evaluated, and the results showed that it possessed homogeneous macroporous morphology, high permeability, and a strong EOF throughout a wide pH range from 2.7 to 11.2. On the constructed monolith, anilines and phenols were effectively separated by CEC, and 2-aminophenol demonstrated the best theoretical efficiency of 470,000 N m^{-1}. Reversed-phase and cation-exchange interactions served as the basic foundation for the retention mechanisms. In order to create organic-silica hybrid monoliths, this work illustrated a novel technique for directly integrating modified nanoparticle monomers.

8. Challenges and future trends

There is still a need for additional work to design distinct inventive NP-based monoliths for various particular applications. The majority of pertinent studies in this sector are, however, still in their infancy, and study in this area has only lately begun, thus there are still many restrictions and difficulties. The issue with characterization is the first worry. It will take further work to fully characterize the substrate, including its chemistry, functional surface, and stability. Because most separation mechanisms are still not fully known, a further obstacle lies in the extensive research in theoretical studies. Furthermore, there is still significant work to be done to ensure their reliable and repeatable production in actual applications, as opposed to only for standard separations. Studies on NPs are still ongoing, with a focus on how they can be used in monolithic matrices for chromatographic separations. These studies aim to control the selectivity of the separations as well as the overall performance of the chromatography by designing and synthesizing highly selective interaction sites on the surfaces of NPs and functionalizing monolithic matrices with a wider variety of NPs bearing different functional groups. In-depth research should also be done on the creation of new functionalization techniques that enable the encapsulation of a greater number of NPs in the polymerization mixture. Additionally, it is important to note that research on innovative organic–inorganic hybrid monolithic columns containing NPs is still in its infancy. Instead, the majority of recent investigations have mostly focused on porous polymer monolithic stationary phase as the matrix for functionalization with NPs. Further research into these NP-modified hybrid materials is therefore necessary in order to uncover novel solutions to some long-standing issues and provide fresh ideas for expanding the extensive application of NPs in chromatography.

Significant advancements have also been made in the functional applications, controlled synthesis, and structural design of MOFs. However, their lackluster stability and exorbitant price restrict their usefulness. MOFs with porous or nonporous silicas can be combined to improve stability and characteristics, and the resulting composites can be used for a variety of applications. Another obstacle is the lack of commercially accessible MOF-based materials for dynamic LC separations, but given how quickly this subject is developing and how many studies have been done in this area, this may not be too far off from becoming a reality. Industry demands that stationary phases be stable in the presence of both water and organic solvents. Additionally, a greater study is required on MOF properties and the variables affecting them. Future research should concentrate on improving existing models that are applicable to MOF synthesis circumstances and creating computer tools that will help researchers improve the conditions and take other elements into account.

9. Concluding remarks

Due to the distinctive qualities of the porous monolithic method, such as simple production, superior permeability, and quick mass transfer, monolithic matrices have attracted increasing interest in the field of liquid chromatographic separation. Monolithic stationary phases have been extensively used in hyphenated systems coupled with mass spectrometers, miniaturized devices, and quick and high-efficiency one- and multi-dimensional separation systems. The separation of complex biological materials using columns is being revolutionized by new monolithic stationary phase preparation technologies. Direct synthesis is a convenient and versatile route to prepare the monolithic columns for microscale separation. Whereas obtaining desired monoliths with chromatographic properties is not easy to succeed because of the complexity of direct synthesis. Compared to the direct synthesis, the post-modification or post-functionalization of monoliths certainly represents a complementary and flexible technique for the preparation of monolithic stationary allows complete independent control of the porous properties.

For the separation of complicated biological samples, such as complex mixtures of peptides for proteome analysis and/or enantiomers, monolithic columns with longer lengths, smaller inner diameters, and specialized selectivity to peptides or enantiomers will play a crucial role in hyphenated systems. A new field of study in the field of chromatographic separation science has been made possible by the invention of monolithic stationary phases, which is now playing a considerably more significant role in a wide range of application areas. In comparison to other carriers like inorganic porous solids (mesoporous silica) or organic polymers, MOFs or coordination polymers have many advantages for the adsorption and release of biomolecules because of their tunable composition, structure, pore size, and volume, easy functionalization, flexible network, and/or accessible metal sites. By selecting the right metal, linker, and structure, one can alter the biodegradable properties of these materials, causing a degradation in body fluid that can last anywhere from a few minutes to many weeks. Even if actual porosity "BioMOFs" are still hard to come by, endogenous linker-based MOFs are one of them and are of particular interest. Making a bioactive MOF with the drug as the linker and releasing the drug through the degradation of the MOF is an alternative method of releasing large amounts of drugs. You can also use a bioactive metal (Ag, Zn, Ca, Mn, Gd, Fe, etc.) as the inorganic cation to add extra properties like antibacterial activity or imaging properties.

Funding

The authors extend their appreciation to the Deanship of Scientific Research and the Research Center, College of Pharmacy, King Saud University for financial support.

Declaration of competing interest

The authors declare no conflict of interest.

Author details

Mohamed Hefnawy[1*], Ali El Gamal[2] and Manal El-Gendy[1]

1 Department of Pharmaceutical Chemistry, College of Pharmacy, King Saud University, Riyadh, Saudi Arabia

2 Department of Pharmacognosy, College of Pharmacy, King Saud University, Riyadh, Saudi Arabia

*Address all correspondence to: mhefnawy@ksu.edu.sa

References

[1] Giddings JC. Dynamics of Chromatography, Principles and Theory. Boca Raton, Florida, USA: CRC Press; 2019

[2] Senchenkova EM, Tswett M. The Creator of Chromatography: Scientific Council on Adsorption and Chromatography. Moscow, Russia: Russian Academy of Sciences; 2003

[3] Holder R, Bollig M, Anderson DR, Hochmuth JKA. Discussion on transport phenomena and three-way kinetics of monolithic converters. Chemical Engineering Science. 2006;**61**:8010-8027

[4] Minakushi H, Nakanishi K, Soga N, Ishizuka N, Tanaka N. Octadecylsilylated porous silica rods as separation media for reversed-phase liquid chromatography. Analytical Chemistry. 1997;**68**:3498-3501

[5] Webster's New Universal Unabridged Dictionary. The Random House Dictionary of the English Language. New York, USA: Barnes and Noble Books; 1996

[6] Guiochon G. Monolithic columns in high-performance liquid chromatography. Journal of Chromatography. A. 2007;**1168**:33-41

[7] Tanaka N, Kobayashi H, Ishizuka N, Minakuchi H, Nakanishi K, Hosoya K, et al. Monolithic silica columns for high-efficiency chromatographic separations. Journal of Chromatography. A. 2002; **965**:35-49

[8] Gzil P, Vervoort N, Baron G, Desmet G. A correlation for the pressure drop in monolithic silica columns. Analytical Chemistry. 2003;**75**:843-850

[9] Hjerten S, Liao JL, Zhang R. High-performance liquid chromatography on continuous polymer beds. Journal of Chromatography. A. 1989;**473**:273-275

[10] Svec F, Frechet JMJ. Continuous rods of macroporous polymer as high-performance liquid chromatography separation media. Analytical Chemistry. 1992;**64**:820-822

[11] Minakuchi H, Nakanishi K, Soga N, Ishizuka T, Tanaka N, N. Octadecylsilylated porous silica rods as separation media for reversed-phase liquid chromatography. Analytical Chemistry. 1996;**68**:3498-3501

[12] Ishizuka N, Minakuchi H, Nakanishi K, Soga N, Nagayama H, Hosoya K, et al. Performance of a monolithic silica column in a capillary under pressure-driven and electrodriven conditions. Analytical Chemistry. 2000;**72**: 1275-1280

[13] Fouad D, Ibrahim FG, Adly A. Ghanem, An insight into chiral monolithic stationary phases for enantioselective HPLC applications. Journal of Separation Science. 2019;**42**: 2303-2340

[14] Liu X, Zhu Q, Chen H, Zhou L, Dang X, Huang J. Preparation of 2,4-dichlorophenoxyacetic acid imprinted organic–inorganic hybrid monolithic column and application to selective solid-phase microextraction. Journal of Chromatography B. 2014;**951–952**:32-37

[15] Ro KW, Nayalk R, Knapp DR. Monolithic media in microfluidic devices for proteomics. Electrophoresis. 2006;**27**: 3547-3553

[16] Ericson C, Li J, Nakazato K, Hjertén S. Preparation of continuous beds for electrochromatography and reversed-

phase liquid chromatography of low-molecular-mass compounds. Journal of Chromatography. A. 1997;**767**:33-41

[17] Marti N, Quattrini F, Butte A, Morbidelli M. Production of polymeric materials with controlled pore structure: The "reactive gelation" process macromol. Materials Engineering. 2005; **290**:221-229

[18] Marti. Institute for Chemical and Bioengineering. Z¨urich, Switzerland: ETH; 2007

[19] Unger KK. Packings and Stationary Phases in Chromatographic Techniques. New York, NY, USA: M. Dekker; 1990

[20] Diack M, Guiochon G. Adsorption isotherms and overloaded elution profiles of phenyl-n-alkanes on porous carbon in liquid chromatography. Langmuir. 1992;**8**:1587-1593

[21] Liapis AI, McCoy MA. Theory of perfusion chromatography. Journal of Chromatography. 1992;**599**:87-104

[22] Nakanishi K, Soga N. Phase separation in silica sol-gel system containing polyacrylic acid I. Gel formation behavior and effect of solvent composition. Journal of Non-Crystal, Solids. 1992;**139**:1-13

[23] Nakanishi K, Soga N. Phase separation in silica sol-gel system containing polyacrylic acid II. Effects of molecular weight and temperature. Journal of Non-Crystalline Solids. 1992; **139**:14-24

[24] Nakanishi K, Minakuchi H, Soga N, Ishizuka N, Tanaka N. Effect of skeleton size on the performance of octadecylsilylated continuous porous silica columns in reversed-phase liquid chromatography. Journal of Chromatography. A. 1997;**762**:135-146

[25] Tanaka N, Ishizuka N, Hosoya K, Kimata K, Minakuchi H, Nakanishi K, et al. Monolithic silicas in separation science: Concepts and Syntheses. Kuromatogurafi. 1993;**14**:50-56

[26] Cabrera K, Wieland G, Lubda D, Nakanishi K, Soga N, Minakuchi H, et al. Monolithic materials: Preparation, properties and applications. Trend. Analytical Chemistry. 1998;**17**:50-59

[27] Cabrera K. Applications of silica-based monolithic HPLC columns. Journal of Separation Science. 2004;**27**:843-849

[28] Leinweber FC, Lubda D, Cabrera K, Tallarek U. Characterization of silica-based monoliths with bimodal pore size distribution. Analytical Chemistry. 2002; **74**:2470-2477

[29] Nakanishi K, Soga N. Phase Separation in Gelling Silica–Organic Polymer Solution: Systems Containing Poly (sodium styrenesulfonate). Journal of the American Ceramic Society. 1991; **74**:2518-2530

[30] Ishizuka N, Minakuchi H, Nakanishi K, Soga N, Tanaka N. Designing monolithic double-pore silica for high-speed liquid chromatography. Journal of Chromatography. A. 1998;**797**:133-137

[31] Siouffi AM. Silica gel-based monoliths prepared by the sol–gel method: Facts and figures. Journal of Chromatography. A. 2003;**1000**:801-818

[32] Minakuchi H, Nakanishi K. Double pore silica gel monolith applied to liquid chromatography. Journal of Sol–Gel Science Technology. 1997;**8**:547-552

[33] Hara T, Izumi Y, Hata K, Baron GV, Bamba T, Desmet G. Performance of small-domain monolithic silica columns in nano-liquid chromatography and comparison with commercial packed bed

columns with 2 µm particles. Journal of Chromatography. A. 2020;**762**:135-146

[34] Pesek JJ, Matyska MT, Larrabee S, S. HPLC retention behavior on hydride-based stationary phases. Journal of Separation Science. 2007;**30**:637-647

[35] Jiang ZJ, Smith NW, Ferguson PD, Taylor MR. Preparation and characterization of long alkyl chain methacrylate-based monolithic column for capillary chromatography. Journal of Biochemical and Biophysical Methods. 2006;**70**:39-45

[36] Huang H, Lin Z, Lin Y, Sun X, Xie Y, Zhang L, et al. Preparation and evaluation of poly(4-vinylphenylboronic acid-co-pentaerythritol triacrylate) monolithic column for capillary liquid chromatography of small molecules and proteins. Journal of Chromatography. A. 2012;**1251**:82-90

[37] Jiang ZJ, Smith NW, Ferguson PD, Taylor MR. Novel highly hydrophilic zwitterionic monolithic column for hydrophilic interaction chromatography. Journal of Separation Science. 2009;**32**: 2544-2555

[38] Jiang ZJJ, Reilly J, Everatt B, Smith NM. Novel zwitterionic polyphosphorylcholine monolithic column for hydrophilic interaction chromatography. Journal of Chromatography. A. 2009;**1216**: 2439-2448

[39] Alpert AJ. Electrostatic repulsion hydrophilic interaction chromatography for isocratic separation of charged solutes and selective isolation of phosphopeptides. Analytical Chemistry. 2008;**80**:62-76

[40] Ibrahim MEA, Zhou T, Lucy CA. Agglomerated silica monolithic column for hydrophilic interaction LC. Journal of Separation Science. 2010;**33**:773-778

[41] Hoegger D, Freitag R. Acrylamide-based monoliths as robust stationary phases for capillary electrochromatography. Journal of Chromatography. A. 2001;**914**:211-222

[42] Maruska C, Ericson A, Vegvari S. (Normal-phase) capillary chromatography using acrylic polymer-based continuous beds. Journal of Chromatography. A. 1999;**837**:25-33

[43] Hoegger D, Freitag R. Investigation of mixed-mode monolithic stationary phases for the analysis of charged amino acids and peptides by capillary electrochromatography. Journal of Chromatography. A. 2003;**1004**:195-208

[44] Saito H, Nakanishi K, Hirao K, Jinnai H. Mutual consistency between simulated and measured pressure drops in silica monoliths based on geometrical parameters obtained by three-dimensional laser scanning confocal microscope observations. Journal of Chromatography. A. 2006;**1119**: 95-104

[45] Hahn R, Panzer M, Hansen E. Mass transfer properties of monoliths. Separation Science & Technology. 2002; **37**:1545-1565

[46] Altmler S, Cabrera K. Structure and performance of silica-based monolithic HPLC columns. Journal of Separation Science. 2008;**31**:2551-2559

[47] Fujimoto C. Charged polyacrylamide gels for capillary Electrochromatographic separations of uncharged, low molecular weight compounds. Analytical Chemistry. 1995;**67**:2050-2053

[48] Palm MV. Novotny, macroporous polyacrylamide/poly(ethylene glycol)

matrixes as stationary phases in capillary electrochromatography. Analytical Chemistry. 1997;**69**:4499-4507

[49] Peters EC, Petro M, Svec F, Frechet JMJ. Molded rigid polymer monoliths as separation media for capillary electrochromatography. Analytical Chemistry. 1997;**69**:3646-3649

[50] Peters EC, Petro M, Svec F, Frechet JMJ. Molded rigid polymer monoliths as separation media for capillary electrochromatography. 1. Fine control of porous properties and surface chemistry. Analytical Chemistry. 1998; **70**:2288-2296

[51] Gusev X, Huang C. Capillary columns with in situ formed porous monolithic packing for micro high-performance liquid chromatography and capillary electrochromatography. Journal of Chromatography. A. 1999;**855**:273-290

[52] Wang QC, Svec F, Frechet JMJ. Macroporous polymeric stationary-phase rod as continuous separation medium for reversed-phase chromatography. Analytical Chemistry. 1993;**65**:2243-2248

[53] Malik. Advances in sol-gel based columns for capillary electrochromatography: Sol-gel open-tubular columns. Electrophoresis. 2002; **23**:3973-3992

[54] Minakuchi H, Nakanishi K, Soga N, Ishizuka N, Tanaka N. Effect of domain size on the performance of octadecylsilylated continuous porous silica columns in reversed-phase liquid chromatography. Journal of Chromatography A. 1998;**797**:121-131

[55] Svec F. Preparation and HPLC applications of rigid macroporous organic polymer monoliths. Journal of Separation Science. 2004;**27**:747-766

[56] Wang QC, Svec F, Frechet JMJ. Reversed-phase chromatography of small molecules and peptides on a continuous rod of macroporous poly (styrene-co-divinylbenzene). Journal of Chromatography. A. 1994;**669**:230-235

[57] Peters EC, Petro M, Svec F, Frechet JMJ. Molded rigid polymer monoliths as separation media for capillary electrochromatography. 2. Effect of chromatographic conditions on the separation. Analytical Chemistry. 1998; **70**:2296-2302

[58] Huang XA, Zhang S, Schultz GA, Henion J. Surface-alkylated polystyrene monolithic columns for peptide analysis in capillary liquid chromatography–electrospray ionization mass spectrometry. Analytical Chemistry. 2002;**74**:2336-2344

[59] Novak BM, Grubbs RH. Catalytic organometallic chemistry in water: The aqueous ring-opening metathesis polymerization of 7-oxanorbornene derivatives. Journal of the American Chemical Society. 1988;**110**:7542-7543

[60] Rieux L, Niederlander H, Verpoorte E, Bischoff R. Silica monolithic columns: Synthesis, characterisation and applications to the analysis of biological molecules. Journal of Separation Science. 2005;**28**:1628-1641

[61] Wu NJ, Thompson R. Fast and efficient separations using reversed phase liquid chromatography. Journal of Liquid Chromatography Related Technology. 2006;**29**:949-988

[62] Xie S, Allington RW, Svec F, Frechet JMJ. Rapid reversed-phase separation of proteins and peptides using optimized 'moulded' monolithic poly(styrene–co-divinylbenzene) columns. Journal of Chromatography. A. 1999;**865**:169-174

[63] Svec F, Peters EC, Sykora D, Yu G, Frechet JMJ. Porous polymer monoliths: An alternative to classical beads. Journal of Higher Research Chromatography. 2000;**23**:3-11

[64] Bisjak CP, Trojer L, Lubbad SH, Wieder W, Bonn GK. Influence of different polymerisation parameters on the separation efficiency of monolithic poly(phenyl acrylate-co-1,4-phenylene diacrylate) capillary columns. Journal of Chromatography. A. 2007; **1154**:269-276

[65] Siouffi AM. About the C term in the van Deemter's equation of plate height in monoliths. Journal of Chromatography. A. 2006;**1126**:86-94

[66] Ou J, Liu Z, Wang H, Lin H, Dong J, Zou H. Recent development of hybrid organic-silica monolithic columns in CEC and capillary LC. Electrophoresis. 2014;**36**:62-76

[67] Urban J, Jandera P, Langmaier P. Effects of functional monomers on retention behavior of small and large molecules in monolithic capillary columns at isocratic and gradient conditions. Journal of Separation Science. 2011;**34**:2054-2062

[68] Puangpila C, Nhujak T, El Rassi Z. Investigation of neutral monolithic capillary columns with varying n-alkyl chain lengths in capillary electrochromatography. Electrophoresis. 2012;**33**:1431-1442

[69] Okanda FM, El Rassi Z. Capillary electrochromatography with monolithic stationary phases. 4. Preparation of neutral stearyl – Acrylate monoliths and their evaluation in capillary electrochromatography of neutral and charged small species as well as peptides and proteins. Electrophoresis. 2005;**26**: 1988-1995

[70] Karenga S, El Rassi Z. Controlling retention, selectivity and magnitude of EOF by segmented monolithic columns consisting of octadecyl and naphthyl monolithic segments – Applications to RP-CEC of both neutral and charged solutes. Electrophoresis. 2011;**32**:90-101

[71] Karenga S, El Rassi Z. Neutral octadecyl monolith for reversed phase capillary electrochromatography of a wide range of solutes. Journal of Separation Science. 2008;**31**:2677-2685

[72] Neequaye T, El Rassi Z. Poly (carboxyethyl acrylate-co-ethylene glycol dimethacrylate) precursor monolith with bonded octadecyl ligands for use in reversed-phase capillary electrochromatography. Electrophoresis. 2021;**42**:2656-2663

[73] Duan Q, Liu C, Liu Z, Zhou Z, Chen W, Wang Q, et al. Preparation and evaluation of a novel monolithic column containing double octadecyl chains for reverse-phase micro high performance liquid chromatography. Journal of Chromatography. A. 2014;**1345**:174-181

[74] Ziang J, Smith NW, Ferguson PD, Taylor MR. Preparation and characterization of long alkyl chain methacrylate-based monolithic column for capillary chromatography. Journal of Biochemical and Biophysical Method. 2007;**70**:39-45

[75] Strege MA. Hydrophilic interaction chromatography–electrospray mass spectrometry analysis of polar compounds for natural product drug discovery. Analytical Chemistry. 1998; **70**:2439-2445

[76] Chen X, Tolley HD, Lee ML. Preparation of zwitterionic polymeric monolithic columns for hydrophilic interaction capillary liquid

chromatography. Journal of Separation Science. 2011;**34**:2088-2096

[77] Buszewski B, Noga S. Hydrophilic interaction liquid chromatography (HILIC)—A powerful separation technique. Analytical and Bioanalytical Chemistry. 2012;**402**:231-247

[78] Cheng J, Chen X, Cai Y, He Y, Chen Z, Lin Z, et al. Preparation and evaluation of a hydrophilic poly(2-hydroxyethyl methacrylate-co-N,N′-methylene bisacrylamide) monolithic column for pressurized capillary electrochromatography. Electrophoresis. 2013;**34**:1189-1196

[79] Lin X, Li Y, Xu D, Yang C, Xie Z. Rapid capillary electrochromatographic profiling of phytohormones on a hydrophilic interaction/strong anion-exchange mixed-mode monolith. Analyst. 2013;**138**:635-641

[80] Lin X, Feng S, Jia W, Ding K, Xie Z. Vinylbenzyl quaternary ammonium-based polymeric monolith with hydrophilic interaction/strong anion exchange mixed-mode for pressurized capillary electrochromatography. Journal of Chromatography A. 2013;**1316**: 104-111

[81] Gunasena DN, El Rassi Z. Neutral, charged and stratified polar monoliths for hydrophilic interaction capillary electrochromatography. Journal of Chromatography. A. 2013;**317**:77-84

[82] Yuan G, Peng Y, Liu Z, Yuan G, Peng Y, Liu Z, et al. A facile and efficient strategy to enhance hydrophilicity of zwitterionic sulfoalkylbetaine type monoliths. Journal of Chromatography. A. 2013;**1301**:88-97

[83] Lin Z, Huang H, Sun X, Lin Y, Zhang L, Chen G. Monolithic column based on a poly(glycidyl methacrylate-co-4-vinylphenylboronic acid-co-ethylene dimethacrylate) copolymer for capillary liquid chromatography of small molecules and proteins. Journal of Chromatography. A. 2012;**1246**:90-97

[84] Chen M-L, Li L-M, Yuan B-F, Ma Q, Feng Y-Q. Preparation and characterization of methacrylate-based monolith for capillary hydrophilic interaction chromatography. Journal of Chromatography. A. 2012;**1230**:54-60

[85] Tian T, Zeng Z, Vulpe D, Casco ME, Divitini G, Midgley PA, et al. A sol-gel monolithic metal-organic framework with enhanced methane uptake. Nature Materials. 2018;**17**:174-179

[86] Jeremias F, Henninger S, Janiak C. Ambient pressure synthesis of MIL-100 (Fe) MOF from homogeneous solution using a redox pathway. Dalton Transactions. 2016;**45**:8637-8645

[87] An J, Geib SJ, Rosi SNL. Cation-triggered drug release from a porous zinc–Adeninate metal–organic framework. Journal of the American Chemical Society. 2009;**131**:8376-8377

[88] Horcajada P, Gref R, Baati A, Allan PK, Maurin G, Couvreur P, et al. Metal–Organic frameworks in biomedicine. Chemical Reviews. 2012;**112**:1232-1268

[89] Dietzel PDC, Blom R, Fjellvag H. Base-induced formation of two magnesium metal-organic framework compounds with a bifunctional Tetratopic ligand. European Journal of Inorganic Chemistry. 2008;**23**:3624-3632

[90] Vimont A, Goupil J-M, Lavalley J-C, Daturi M, Surbl S, Serre C, et al. Investigation of acid sites in a zeotypic giant pores chromium(III) carboxylate. Journal of the American Chemical Society. 2006;**128**:3218-3227

[91] McKinlay AC, Xiao B, Wragg DS, Wheatley PS, Megso IL, Morris E. Exceptional behavior over the whole adsorption–storage–delivery cycle for NO in porous metal organic frameworks. Journal of the American Chemical Society. 2008;**130**:10440-10444

[92] Horcajada P, Serre C, Vallet-Regí M, Sebban K, Taulelle F, Ferey G. Metal-organic frameworks as efficient materials for drug delivery. Angewandte Chemie International Edition. 2006;**45**: 5974-5978

[93] Li W, Jin L, Zhu N, Hou X, Deng F, Sun H. Structure of colloidal bismuth subcitrate (CBS) in dilute HCl: Unique assembly of bismuth citrate Dinuclear units ([Bi(cit)2Bi]2-). Journal of the American Chemical Society. 2003;**125**: 12408-12409

[94] Rood JA, Noll BC, Henderson KW. Synthesis, structural characterization, gas sorption and guest-exchange studies of the lightweight, porous metal–organic framework α-[$Mg_3(O_2CH)_6$]. Inorganic Chemistry. 2006;**45**:5521-5528

[95] Ingleson MJ, Barrio JP, Bacsa J, Dickinson C, Park H, Rosseinsky M. Generation of a solid Brønsted acid site in a chiral framework. Chemical Communications. 2008;**2008**:1287-1289

[96] Rabone J, Yue Y-F, Chong SY, Stylianou KC, Bacsa J, Bradshaw D, et al. An adaptable peptide-based porous material. Science. 2010;**329**:1053-1057

[97] Weber R, Bergerhoff GZ, Kristallogr Z. Crystal structure of iron(III) oxonium 3,4,5-trihydroxybenzoate, ($C_7H_2O_5$)Fe (H_3O). Crystals Materials. 1991;**195**:87-88

[98] Serre C, Millange F, Surble S, Ferey F. A route to the synthesis of trivalent transition-metal porous carboxylates with trimeric secondary building units. Angewandte Chemie, International Edition. 2004;**43**:6286-6289

[99] Serre C, Surble S, Mellot-Draznieks C, Filinchuk Y, Ferey G. Evidence of flexibility in the nanoporous iron(iii) carboxylate MIL-89. Dalton Transactions. 2008;**2008**:5462-5464

[100] Eddaoudi M, Kim J, Rosi N, Vodak D, Wachter J, O'Keeffe M, et al. Systematic design of pore size and functionality in isoreticular MOFs and their application in methane storage. Science. 2002;**295**:469-472

[101] Li H, Eddaoudi M, O'Keeffe M, Yaghi M. Design and synthesis of an exceptionally stable and highly porous metal-organic framework. Nature. 1999; **402**.276-283

[102] Devic T, Horcajada P, Serre C, Moulin B, Leclerc H, Heurtaux D, et al. Functionalization in flexible porous solids: Effects on the pore opening and the host–guest interactions. Journal of the American Chemical Society. 2010; **132**:1127-1136

[103] Horcajada P, Salles F, Wuttke S, Devic T, Heutaux D, Maurin A, et al. How Linker's modification controls swelling properties of highly flexible Iron (III) dicarboxylates MIL-88. Journal of the American Chemical Society. 2011; **133**:17839-17847

[104] Hermes S, Witte T, Hikov T, Zacher D, S. Bahnm€uller, G. Langstein, K. Huber, R.A. Fischer, rapping metal-organic framework Nano-crystals: An in-situ time-resolved light scattering study on the crystal growth of MOF-5 in solution. Journal of the American Chemical Society. 2007;**129**:5324-5325

[105] Horcajada P, Chalati T, Serre C, Gillet B, Sebrie C, Baati T, et al. Porous

metal–organic-framework nanoscale carriers as a potential platform for drug delivery and imaging. Nature Materials. 2010;**9**:172181

[106] Horcajada P, Serre C, Grosso D, Boissiere C, Perruchas S, Sanchez C, et al. Colloidal route for preparing optical thin films of nanoporous metal–organic frameworks. Advanced Materials. 2009; **21**:1931-1935

[107] Chalati T, Horcajada P, Gref R, Couvreur P, Serre C. Optimization of the synthesis of MOF nanoparticles made of flexible porous iron fumarate MIL-88A. Journal of Materials Chemistry. 2011;**21**: 2220-2227

[108] Cho W, Lee HJ, Oh HM. Growth controlled formation of porous coordination polymer particles. Journal of the American Chemical Society. 2008; **130**:16943-16946

[109] Tsuruoka T, Furukawa S, Takashima Y, Yoshida K, Isoda S, Kitagawa S. Nanoporous nanorods fabricated by coordination modulation and oriented attachment growth. Angewandte Chemie, International Edition. 2009;**48**:4739-4743

[110] Kerbellec N, Catala L, Daiguebonne C, Gloter A, Stephan O, Bunzli B, et al. Luminescent coordination nanoparticles. New Journal of Chemistry. 2008;**32**: 584-587

[111] Daiguebonne C, Kerbellec N, Guillou O, Bcunzli J, Gumy F, Catala L, et al. Structural and luminescent properties of micro- and nanosized particles of lanthanide terephthalate coordination polymers. Inorganic Chemistry. 2008;**47**:3700-3708

[112] Rieter WJ, Taylor KML, An H, Lin W. Nanoscale metal–organic frameworks as potential multimodal contrast enhancing agents. Journal of the American Chemical Society. 2006;**128**: 9024-9025

[113] Taylor KML, Jin A, Lin W. Surfactant-Assisted Synthesis of Nanoscale Gadolinium Metal–Organic Frameworks for Potential Multimodal Imaging. Angewandte Chemie International Edition. 2008;**47**:7722-7725

[114] Rieter WJ, Pott KM, Taylor KML, Lin W. Nanoscale coordination polymers for platinum-based anticancer drug delivery. Journal of the American Chemical Society. 2008;**35**:11584-11585

[115] Taylor KML, Rieter WJ, Lin W. Manganese-based nanoscale metal–organic frameworks for magnetic resonance imaging. Journal of the American Chemical Society. 2008;**130**: 14358-13459

[116] Fillion H, Luche J. In Synthetic Organic Sonochemistry Luche. New York, USA: Plenum Press; 1998

[117] Suslick S. Ultrasound: Its Chemical, Physical, and Biological Effects. Weinheim, Germany: VCH; 1988

[118] Schlesinger M, Schulze S, Hietschold M, Mehring M. Evaluation of synthetic methods for microporous metal–organic frameworks exemplified by the competitive formation of [Cu2 (btc)3(H2O)3] and [Cu2(btc)(OH) (H2O)]. Microporous and Mesoporous Materials. 2010;**132**:121-127

[119] Seo Y, Yoon J, Lee U, Hwang Y, Jun C, Chang J. Formation of a nanohybrid composite between mesostructured cellular silica foam and icroporous copper trimesate. Microporous and Mesoporous Materials. 2012;**155**:75-81

[120] Qiu L-G, Li Z-Q, Wu Y, Wang W, Xu T, Jiang S. Facile synthesis of

nanocrystals of a microporous metal–organic framework by an ultrasonic method and selective sensing of organoamines. Chemical Communication. 2008:3642-3644

[121] Hwang YK, Chang J, Kwon Y, Park S. Microwave synthesis of cubic mesoporous silica SBA-16. Microporous and Mesoporous Materials. 2004;**68**: 21-27

[122] Ni Z, Maesl R. Rapid production of metal—organic frameworks via microwave-assisted solvothermal synthesis. Journal of the American Chemical Society. 2006;**128**:12394-12395

[123] Jhung SH, Lee J-H, Yoon JW, Serre C, Ferey G, Chang JS. Microwave synthesis of chromium terephthalate MIL-101 and its benzene sorption ability. Advanced Materials. 2007;**19**:121-127

[124] Demessence A, Horcajada P, Serre C, Boissiere C, Grosso D, Sanchez C. Ferey, elaboration and properties of hierarchically structured optical thin films of MIL-101(Cr). Chemical Communications. 2009;**23**:7149-7155

[125] Sefcik J, McCormick AV. Kinetic and thermodynamic issues in the early stages of sol-gel processes using silicon alkoxides. Catalysis Today. 1997;**35**:205-223

[126] Kajihara K, Goto M. Cosolvent-free synthesis of macroporous silica gels and monolithic silica glasses from tetraalkoxysilane–water binary systems: Comparison between tetramethoxysilane and tetraethoxysilane. Journal of Sol-Gel Science and Technology. 2022;**104**: 497-502

[127] Dulay MT, Kulkarni RP, Zare RN. Preparation and characterization of monolithic porous capillary columns loaded with chromatographic particles. Analytical Chemistry. 1998;**70**:5103-5107

[128] Bakry R, Stoggl WM, Hochleitner EO, Stecher G, Huck CW, Bonn GK. Silica particles encapsulated poly (styrene-divinylbenzene) monolithic stationary phases for micro-high performance liquid chromatography. Journal of Chromatography. A. 2006; **1132**:183-189

[129] Hench LL, West JK. The sol-gel process. Chemical Reviews. 1990;**90**: 33-72

[130] Xu L, Lee HK. Preparation, characterization and analytical application of a hybrid organic–inorganic silica-based monolith. Journal of Chromatography. A. 2008;**1195**(1195): 78-84

[131] Chen M, Zhang J, Zhang Z, Yuan B, Yu Q, Feng Y. Facile preparation of organic-silica hybrid monolith for capillary hydrophilic liquid chromatography based on "thiol-ene" click chemistry. Journal of Chromatography. A. 2013;**1284**:118-125

[132] Colon H, Zhang X, Murphy JK, Rivera JG, Colon LA. Allyl-functionalized hybrid silica monoliths. Chemical Communications. 2005;**2005**: 2826-2828

[133] Yan L, Zhang Q, Feng Y, Zhang W, Li T, Zhang Y. Octyl-functionalized hybrid silica monolithic column for reversed-phase capillary electrochromatography. Journal of Chromatography. A. 2006;**1121**:92-98

[134] Hayes JD, Malik A. Sol—gel monolithic columns with reversed electroosmotic flow for capillary Electrochromatograph. Analytical Chemistry. 2000;**72**:4090-4099

[135] Laschober S, Sulyok M, Rosenberg E. Tailoring the macroporous structure of monolithic silica-based capillary

columns with potential for liquid chromatography. Journal of Chromatography. A. 2007;**1144**:55-62

[136] Dunn B, Zink JI. Sol–gel chemistry and materials. Chemical Research. 2007; **40**:729-747

[137] Staniak M, Wójciak M, Sowa I, Tyszczuk-Rotko K, Strzemski M, Dresler S, et al. Silica-based monolithic columns as a tool in HPLC—An overview of application in analysis of active compounds in biological samples. Molecules. 2020;**25**:3149-3170

[138] Jiang Z, Smith N, Liu Z. Preparation and application of hydrophilic monolithic columns. Journal of Chromatography. A. 2011;**1218**: 2350-2361

[139] Nakanishi K, Soga N. Inorganic porous column, Japan Patent 5-1993, 200, 392

[140] Pack BW, Risley DS. Evaluation of a monolithic silica column operated in the hydrophilic interaction chromatography mode with evaporative light scattering detection for the separation and detection of counter-ions. Journal of Chromatography. A. 2005; **1073**:269-275

[141] Puy G, Demesmay C, Rocca JL, Iapichella J, Galarneau A, Brunel D. Electrochromatographic behavior of silica monolithic capillaries of different skeleton sizes synthesized with a simplified and shortened sol-gel procedure. Electrophoresis. 2006;**27**: 3971-3980

[142] Puy G, Roux R, Demesmay C, Rocca JL, Iapichella J, Galameau A, et al. Influence of the hydrothermal treatment on the chromatographic properties of monolithic silica capillaries for nano-liquid chromatography or capillary electrochromatography. Journal of Chromatography. A. 2007;**1160**:150-159

[143] Liu H, Yu H, Jin P, Jiang M, Zhu G, Duan Y, et al. Preparation of mesoporous silica materials functionalized with various amino-ligands and investigation of adsorption performances on aromatic acids. Chemical Engineering Journal. 2020;**379**:122405

[144] Ikegami T, Fujita H, Horie K, Hosoya K, Tanaka N. HILIC mode separation of polar compounds by monolithic silica capillary columns coated with polyacrylamide. Analytical and Bioanalytical Chemistry. 2006;**386**: 578-885

[145] Alaerts L, Kirschhock CE, Maes M, Van Der Veen MA, Finsy V, Depla A, et al. Selective adsorption and separation of xylene isomers and ethylbenzene with the microporous vanadium (IV) terephthalate MIL-47. Angewandte Chemie, International Edition. 2007;**46**: 4293-4297

[146] Yang CX, Chen YJ, Wang HF, Yan XP. High-performance separation of fullerenes on metal–organic framework MIL-101 (Cr). Chemistry - A European Journal. 2011;**17**:11734-11737

[147] Fu Y-Y, Yang C-X, Yan X-P. Incorporation of metal–organic framework UiO-66 into porous polymer monoliths to enhance the liquid chromatographic separation of small molecules. Chemical Communications. 2013;**49**:7162-7164

[148] Yang F, Yang C-X, Yan X-P. Post-synthetic modification of MIL-101 (Cr) with pyridine for high-performance liquid chromatographic separation of tocopherols. Talanta. 2015;**137**:136-142

[149] Li L, Sun F, Jia J, Borjigin T, Zhu G. Growth of large single MOF crystals and

effective separation of organic dyes. CrystEngComm. 2013;**15**:4094-4098

[150] Zhao S-N, Krishnaraj C, Jena HS, Poelman D, Van Der Voort P. An anionic metal-organic framework as a platform for charge-and size-dependent selective removal of cationic dyes. Dyes and Pigments. 2018;**156**:332-337

[151] Zhao W-W, Zhang C-Y, Yan Z-G, Bai L-P, Wang X, Huang H, et al. Separations of substituted benzenes and polycyclic aromatic hydrocarbons using normal-and reverse-phase high performance liquid chromatography with UiO-66 as the stationary phase. Journal of Chromatography. A. 2014; **1370**:121-128

[152] Fu YY, Yang CX, Yan XP. Fabrication of ZIF-8@ SiO2 core–shell microspheres as the stationary phase for high-performance liquid chromatography. Chemical European Journal. 2013;**19**:13484-13491

[153] Ahmad R, Wong-Foy AG, Matzger AJ. Microporous coordination polymers as selective sorbents for liquid chromatography. Langmuir. 2009;**25**: 11977-11979

[154] Centrone A, Santiso EE, Hatton TA. Separation of chemical reaction intermediates by metal–organic frameworks. Small. 2011;7:2356-2364

[155] Yan Z, Zhang W, Gao J, Lin Y, Li J, Lin Z, et al. Reverse-phase high performance liquid chromatography separation of positional isomers on a MIL-53 (Fe) packed column. RSC Advances. 2015;**5**:40094-40102

[156] Liu S-S, Yang C-X, Wang S-W, Yan X-P. Metal–organic frameworks for reverse-phase high-performance liquid chromatography. Analyst. 2012;**137**: 816-818

[157] Yang C-X, Yan X-P. Metal–organic framework MIL-101 (Cr) for high-performance liquid chromatographic separation of substituted aromatics. Analytical Chemistry. 2011;**83**:7144-7150

[158] Maes M, Vermoortele F, Alaerts L, Couck S, Kirschhock CE, Denayer JF, et al. Separation of styrene and ethylbenzene on metal - organic frameworks: Analogous structures with different adsorption mechanisms. Journal of the American Chemical Society. 2010;**132**:15277-15285

[159] Si T, Wang S, Zhang H, Wang L, Lu X, Liang X, et al. Design and evaluation of novel MOF–polymer core–shell composite as mixed-mode stationary phase for high performance liquid chromatography. Microchimica Acta. 2021;**188**:1 9

[160] Ehrling S, Kutzscher C, Freund P, Müller P, Senkovska I, Kaskel S. MOF@ SiO2 core-shell composites as stationary phase in high performance liquid chromatography. Microporous and Mesoporous Materials. 2018;**263**:268-274

[161] Fan F, Lu X, Liang X, Wang L, Guo Y. Preparation of hydrogel nanocomposite functionalized silica microspheres and its application in mixed-mode liquid chromatography. Journal of Chromatography. A. 2022; **1662**:462745

[162] Nuzhdin AL, Dybtsev DN, Bryliakov KP, Talsi EP, Fedin VP. Enantioselective chromatographic resolution and one-pot synthesis of enantiomerically pure sulfoxides over a homochiral Zn - organic framework. Journal of the American Chemical Society. 2007;**129**:12958-12959

[163] Fei Z-X, Zhang M, Zhang J-H, Yuan L-M. Chiral metal–organic framework used as stationary phases for capillary

electrochromatography. Analytica Chimica Acta. 2014;**830**:49-55

[164] Chang C-L, Qi X-Y, Zhang J-W, Qiu Y-M, Li X-J, Wang X, et al. Facile synthesis of magnetic homochiral metal–organic frameworks for "enantioselective fishing". Chemical Communications. 2015;**51**:3566-3569

[165] Corella-Ochoa MN, Tapia JB, Rubin HN, Lillo V, Gonzalez-Cobos J, Núnez-Rico JL, et al. Homochiral metal–organic frameworks for enantioselective separations in liquid chromatography. Journal of the American Chemical Society. 2019;**141**:14306-14316

[166] Yu Y, Xu N, Zhang J, Wang B, Xie S, Yuan L. Chiral metal–organic framework d-his-ZIF-8@ SiO2 core–shell microspheres used for HPLC enantioseparations. ACS Applied Materials & Interfaces. 2020;**12**: 16903-16911

[167] Mao Z, Bao T, Li Z, Chen Z. Ionic liquid-copolymerized monolith incorporated with zeolitic imidazolate framework-8 as stationary phases for enhancing reversed phase selectivity in capillary electrochromatography. Journal of Chromatography. A. 2018;**1578**:99-105

[168] Ye FG, Lu JY, Wang Y, Zhang AZ, Tian JN, Zhao SL. Preparation and characterization of gold nanoparticle-modified silica monolith for capillary electrochromatography. Chinese Journal of Analytical Chemistry. 2011;**39**: 341-345

[169] Grabar KC, Freeman RG, Hommer MB, Natan MJ. Preparation and characterization of Au colloid monolayers. Analytical Chemistry. 1995; **67**:735-743

[170] Zhang A, Ye F, Lu J, Zhao S. Screening α-glucosidase inhibitor from natural products by capillary electrophoresis with immobilised enzyme onto polymer monolith modified by gold nanoparticles. Food Chemistry. 2013;**141**:1854-1859

[171] Lu J, Ye F, Zhang A, Wei Z, Peng Y, Zhao S. Preparation and characterization of silica monolith modified with bovine serum albumin-gold nanoparticles conjugates and its use as chiral stationary phases for capillary electrochromatography. Journal of Separation Science. 2011;**34**:2329-2336

[172] Lv Y, Alejandro FM, Frechet JM, Svec F. Preparation of porous polymer monoliths featuring enhanced surface coverage with gold nanoparticles. Journal of Chromatography A. 2012; **A1261**:121-128

[173] Lv Y, Lin Z, Svec F. Hypercrosslinked large surface area porous polymer monoliths for hydrophilic interaction liquid chromatography of small molecules featuring zwitterionic functionalities attached to gold nanoparticles held in layered structure. Analytical Chemistry. 2012;**84**:8457-8460

[174] Chang LT, Yen CC. Studies on the preparation and properties of conductive polymers. VIII. Use of heat treatment to prepare metallized films from silver chelate of PVA and PAN. Journal of Applied Polymer Science. 1995;**55**: 371-374

[175] Feng Q, Cui F, Kim T, Kim J. Ag-substituted hydroxyapatite coatings with both antimicrobial effects and biocompatibility. Journal of Materials Science Letters. 1999;**18**:559-561

[176] Fritzsche W, Porwol H, Wiegand A, Bornmann S, Kohler J. In-situ formation of Ag-containing nanoparticles in thin

polymer films. Nanostructured Materials. 1998;**10**:89-97

[177] Navarro-Pascual-Ahuir M, Lerma-Garcr MJ, Ler-Ramos MJ, Simm-Alfonso EF, Herrero-Martınez JM. Preparation and evaluation of lauryl methacrylate monoliths with embedded silver nanoparticles for capillary electrochromatography. Electrophoresis. 2013;**34**:925-934

[178] Kucka J, Hruby M, Konak C, Kozempel J, Lebeda O. Astatination of nanoparticles containing silver as possible carriers of 211At. Applied Radiation and Isotopes. 2006;**64**:201-206

[179] Sedlacek O, Kucka J, Svec F, Hruby M. Silver-coated monolithic columns for separation in radiopharmaceutical applications. Journal of Separation Science. 2014;**37**:798-802

[180] Randon J, Guerrin J-F, Rocca J-L. Synthesis of titania monoliths for chromatographic separations. Journal of Chromatography. A. 2008;**1214**: 183-186

[181] Jaoude MA, Randon J. Capillary monolithic titania column for miniaturized liquid chromatography and extraction of organo-phosphorous compounds. Analytical and Bioanalytical Chemistry. 2011;**400**:1241-1249

[182] Wan J, Qian K, Qiao L, Wang Y, Kong J, Yang P, et al. TiO2-modified macroporous silica foams for advanced enrichment of multi-phosphorylated peptides. Chemical European Journal. 2009;**15**:2504-2508

[183] Zhang Y, Chen C, Qin H, Wu R, Zou H. The synthesis of Ti-hexagonal mesoporous silica for selective capture of phosphopeptides. Chemical Communications. 2010;**46**:2271-2273

[184] Wu JH, Li XS, Zhao Y, Zhang W, Guo L, Feng YQ. Application of liquid phase deposited titania nanoparticles on silica spheres to phosphopeptide enrichment and high performance liquid chromatography packings. Journal of Chromatography. A. 2011;**1218**: 2944-2953

[185] Wang ST, Wang MY, Su X, Yuan BF, Feng YQ. Facile preparation of SiO_2/ TiO_2 composite monolithic capillary column and its application in enrichment of Phosphopeptides. Analytical Chemistry. 2012;**84**:7763-7770

[186] Nawrocki J, Rigney M, McCormick A, Carr P. Chemistry of zirconia and its use in chromatography. Journal of Chromatography. A. 1993;**657**:229-282

[187] Nawrocki J, Dunlap C, McCormick A, Carr P, Part I. Chromatography using ultra-stable metal oxide-based stationary phases for HPLC. Journal of Chromatography. A. 2004;**1028**:1-30

[188] Dunlap CJ, Carr PW, McNeff CV, Stoll D. Zirconia stationary phases for extreme separations. Analytical Chemistry. 2001;**73**:598A-607A

[189] Nawrocki J, Dunlap CJ, Carr PW, Blackwell JA. New materials for biotechnology: Chromatographic stationary phases based on zirconia. Biotechnology Progress. 1994;**10**:561-573

[190] Kumar AP, Park JH. Chiral separation of basic compounds on a cellulose 3,5-dimethylphenylcarbamate-coated zirconia monolithin basic eluents by capillary electrochromatography. Journal of Chromatography. A. 2011; **1218**:6548-6553

[191] Kumar AP, Park JH. Fast separations of chiral β-blockers on a cellulose tris(3,5-dimethylphenylcarbamate)-coated zirconia monolithic column by capillary

electrochromatography. Journal of Chromatography. A. 2011;**1218**:5369-5373

[192] Kim M, Park JH. Enantioseparation of chiral acids and bases on aclindamycin phosphate-modified zirconia monolith by capillary electrochromatography. Journal of Chromatography. A. 2012; **1251**:244-248

[193] Hong JS, Park JH. Chiral separation of basic compounds on Sulfated β-cyclodextrin-coated zirconia monolith by capillary electrochromatography. Bulletin of the Korean Chemical Society. 2013;**34**:1809-1814

[194] Lee JM, Jang MD, Park JH. Chiral separation on sulfonated cellulose tris (3,5-dimethylphenylcarbamate)-coated zirconia monolith by capillary electrochromatography. Bulletin of the Korean Chemical Society. 2012;**33**:2651-2656

[195] Liu Y, Chen Y, Yang H, Nie L, Yao S. Cage-like silica nanoparticles-functionalized silica hybrid monolith for high performance capillary electrochromatography via "one-pot" process. Journal of Chromatography. A. 2013;**1283**:132-139

Chapter 4

Perspective Chapter: High-Performance Liquid Chromatography Coupled to Mass Spectrometry – The Advance in Chemical Analysis

Kelly da Silva Bezerra

Abstract

How these techniques coupled, high-performance liquid chromatography and mass spectrometry (LC-MS), revolutionized the field of chemical analysis. For a long time, the field of instrumental chemical analysis was limited, as we could not reliably quantify low-concentration substances in complex matrices. This is certainly one of the main advantages of coupling these techniques. In this perspective chapter, the advantages and limitations of the LC-MS system will be presented, sample preparation for this type of analysis and examples of applications that have had analytical success. It will also be mentioned the analytical advance in relation to miniaturization for lab-in-syringe, lab-on-valve, lab-on-a-chip, organ-on-a-chip, among others.

Keywords: quantitative analysis, LC-MS, mass spectrometry, chemical analysis, miniaturization of LC

1. Introduction

High-performance liquid chromatography (HPLC) methods have been developed for the determination of chemicals at low concentration levels in complex samples. When we go deeper into the analytical instrumentation, we arrive at the application of ultra liquid chromatography (UHPLC) which has had a growing tendency to use this separation technique, guaranteeing faster assays compared to HPLC [1–3].

Among the main detectors used in such applications, stands out the mass spectrometers [4–6]. The application of mass spectrometry in qualitative and quantitative analysis has stood out in recent years, in the identification of analytes in complex mixtures, and especially in the structural characterization of unknown compounds [5, 7]. Thus, mass spectrometry provides accurate information about the compounds of interest. Currently, tandem mass spectrometry (MS/MS) has become indispensable in the characterization of analytes, mainly due to its high sensitivity [7–9].

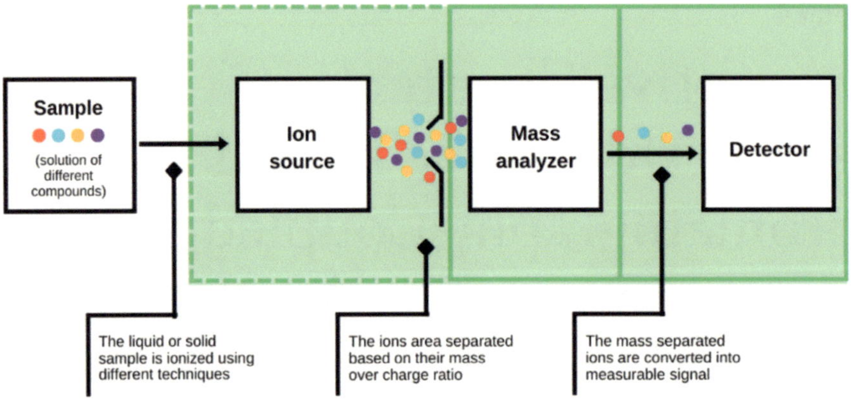

Figure 1.
Representation of the parts of a mass spectrometer: ion source, mass analyzer, and system of detection.

The literature mentions the use of triple quadrupoles, time of flight (TOF), ion trap, and orbitrap (electrostatic with ion trap) as the main analyzers in applications [4, 5, 7]. The most used ionization methods are electrospray (ESI), atmospheric pressure photoionization (APPI) or atmospheric pressure chemical ionization (APCI) [4, 5]. **Figure 1** shows a representation of the parts of a mass spectrometer.

When we talk about the analysis of chemical compounds in complex matrices such as residues, contaminants, adulterants, metabolites, and degradation products, there are limitations that range from the costly preparation of some samples to the difficulty of using the same configuration of the LC-MS/MS for the identification and quantification of several analytes in the same sample [10, 11]. In this progress of analytical science, the more powerful the instruments are, the more complex their operation is, demanding a high level of expertise on the part of the operator.

We also need to mention, inspiring analytical advances that were made in relation to even more efficient strategies, through hyphenated systems, miniaturization for lab-in-syringe, lab-on-valve, lab-on-a-chip, organ-on-a-chip, 3D printed technologies, artificial intelligence, among others [12–16]. In this chapter, the power of hyphenated techniques such as LC-MS in the study and routine application of different analytes in different types of samples and fields of application will be presented.

2. The union of high-performance liquid chromatography (HPLC) with the mass spectrometer (MS)

The current trend in HPLC is to reduce analysis time and increase sample throughput without sacrificing the selectivity factor. The first possibility is to shorten the column length along with reducing the particle diameter. Another important parameter is the temperature, as the separation at a high temperature can reduce the analysis time [1, 2].

HPLC in reversed-phase mode (mainly using C18 columns), often referred to as RPLC, plays a dominant role for the determination of macromolecules, while UHPLC, in the same chromatographic mode, reduces runtime and improves sensitivity [1–3, 17].

Compared with GC, HPLC has some disadvantages in chromatographic performance such as resolution. The recent introduction of UHPLC employs porous particles with internal diameters smaller than 2 μm which, in combination with MS, results in higher peak capacity, better resolution, and higher sensitivity compared to conventional HPLC columns [3, 17].

In parallel, mass spectrometry (MS) is widely recognized as a powerful analytical tool that provides quality quantitative data that may not be readily available by other instrumental techniques [4–6], for example, molecular weight, empirical formula (through precise mass measurement), isotopic ratios, detection of functional groups and other substituents, and structure elucidation, including in some cases, stereochemistry [7–9]. Among other advantages of the LC-MS coupling, we mention the extremely high sensitivity and low sample consumption [4, 8, 9].

LC-MS applications generally employ reversed-phase separations; however, when hydrophilic interaction chromatography (HILIC) and ion pair chromatography (IPC) are applied, compounds of greater polarity can be accommodated, due to the limitations of RPLC [9, 18]. HILIC has an important unique benefit over normal phase LC (NPLC) in that aqueous samples have a much higher compatibility with the mobile phase, simplifying sample introduction. However, the load capacity of the HILIC is generally much less than that of the RPLC. HILIC also tends to generate wider peaks than RPLC, resulting in a lower selectivity factor and greater reliance on the mass spectrometer for peak resolution [4, 8, 9, 18].

With the development of mass spectrometry, a whole new field has opened up within analytical chemistry, providing new opportunities for the analysis of diverse substances. The data acquisition modes in the LC-MS/MS instruments made the chemical profile of a sample accessible, within the limitations of the instrument [8–10].

The main advantages of LC-MS are as follows: the ability to identify unknown compounds from complex matrices (i.e., food, environmental, biological samples); solve co-elution problems, since there are two identification parameters (retention time and m/z fragmentation), making it possible to select and separate the desired transitions even if the compounds are co-eluted in the chromatogram [2, 4, 6, 10].

In coupling LC with sequential mass spectrometry (LC-MS/MS), three atmospheric pressure ionization (API) techniques cover virtually the entire range of compound polarities, namely electrospray ionization (ESI), chemical ionization at atmospheric pressure (APCI), and atmospheric pressure photoionization (APPI) [5–7, 19]. **Figure 2** shows the ionizations sources. However, a single form of ionization cannot cover all types of molecules: polar, neutral, ionic, and non-polar. Thus, the ionization mode can be chosen according to the acidic, neutral, or basic character of the analytes [19].

In some cases, the correct choice of ionization technique and polarity mode (positive or negative) is not so obvious. ESI ionization in positive mode, for example, is the most common mode in LC-MS/MS because it can effectively ionize a wide range of low and medium polarity molecules [4, 7, 19]. Negative ionization provides better results for certain analyte classes (e.g., organic acids or carbohydrates) [19]. APCI, on the other hand, is preferred for more non-polar analytes, for example [4–6, 19]. Thus, tests in all available ionization modes are very important as a first step in method optimization.

The characteristic ions are the same for all API techniques: the base peak of the positive ion mass spectrum in full scan mode is generally the $[M+H]^+$ ion often accompanied by less abundant alkali metal adducts $[M+Na]^+$ and $[M+K]^+$ [5, 19, 20]. On the other hand, the negative ion mode normally shows the peak of the deprotonated molecule $[M-H]^-$ as the base peak [20]. APPI more often shows the formation of M^+ radical molecular ions, unlike the other API techniques. Rarely, adducts with HPLC mobile phase molecules are found as $[M+H+solvent]^+$ in HPLC-MS/MS coupling [5, 20].

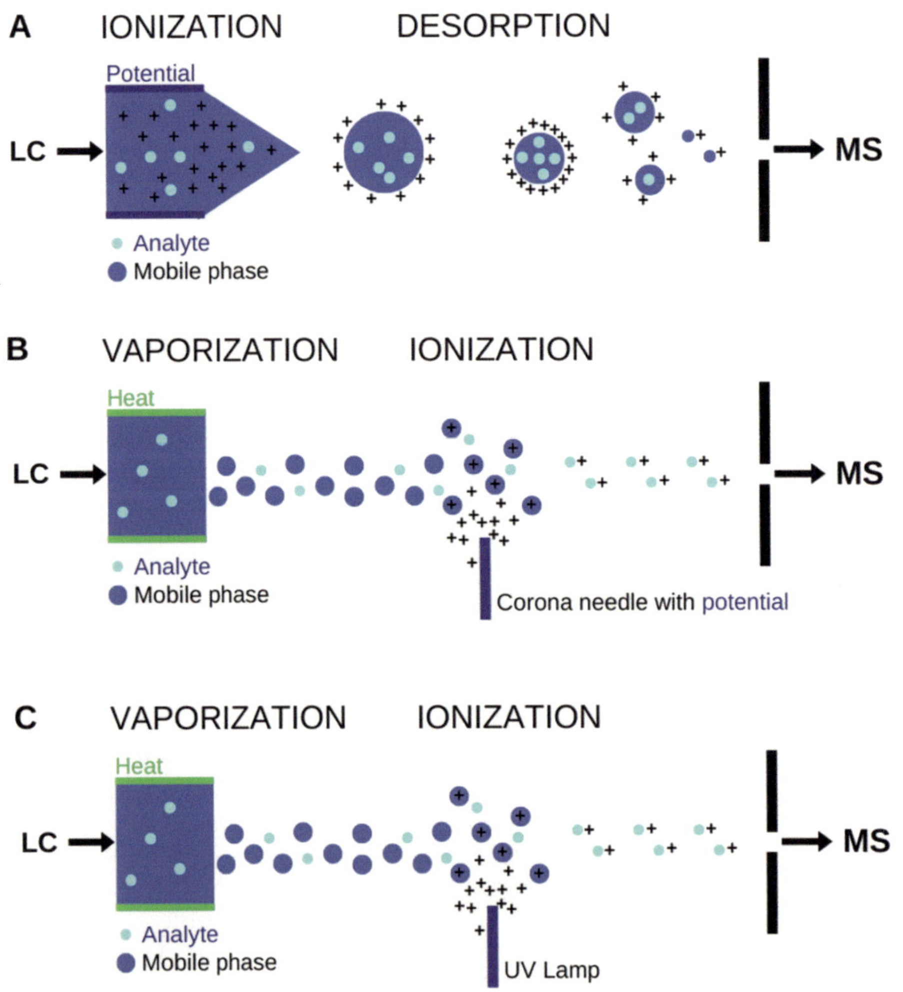

Figure 2.
Representation of the main sources of ionization for mass spectrometry. (A) Electrospray ionization (ESI), (B) chemical ionization at atmospheric pressure (APCI), and (C) atmospheric pressure photoionization (APPI).

Recently, there has also been considerable interest in laser desorption ionization (LDI) techniques such as matrix-assisted laser desorption Ionization (MALDI) and surface-enhanced laser desorption ionization (SELDI). MALDI is the second most important ionization technique in complex applications, but its role in metabolite identification is not as comprehensive [20–22]. Several other ionization modes are constantly being proposed, including direct analysis in real time (DART), extractive electrospray ionization (EESI), laser ablation electrospray ionization (LAESI), and paper spray ionization, among others [21, 22].

The function of a mass analyzer is the separation of ions according to their different mass-to-charge ratios (m/z). The quality of mass separation is characterized by the degree to which close m/z values can be separated by the analyzer. Mass analyzers are classified into low- and high-resolution groups [6–9]. In general, low-resolution mass analyzers (e.g., quadrupoles and ion traps) have resolution in the many thousand range, while high-resolution analyzers should provide a minimum resolution of

15,000, for time-of-flight (TOF)-based analyzers, or more than 50,000, for Fourier transform mass analyzers (FTMS) [21, 22].

The advantage of ion traps is the possibility of studying fragmentation mechanisms [20, 22]. Some papers showed that the selectivity of the peaks and the number of tracer ions detected using the fast separations of UHPLC and TOF/MS were similar to those generated by the conventional methods of HPLC/MS with a reduction of 10 minutes in the analysis time [10, 22]. Fourier transform ion cyclotron resonance mass spectrometry (FTICR) provides the highest resolving power available currently. However, due to its complexity and high cost, FTICR is rarely used in conjunction with LC [10, 20–22].

Sophisticated HPLC or UHPLC systems associated with mass spectrometry (MS) exposed the difficulties associated with the analysis of new analytes in complex samples and analytes at very low concentrations (from nanogram to picogram) [6–9, 20, 21]. The basic incompatibility between LC and MS stems mainly from two factors.

First, the mass spectrometer requires ions in the gas phase at very low pressures compared to the pressures that are applied in HPLC. Another reason is that HPLC employs solutions often containing non-volatile solutes and buffers at atmospheric pressure, at flow rates of 1 mL/min. This flux typically corresponds to about 100

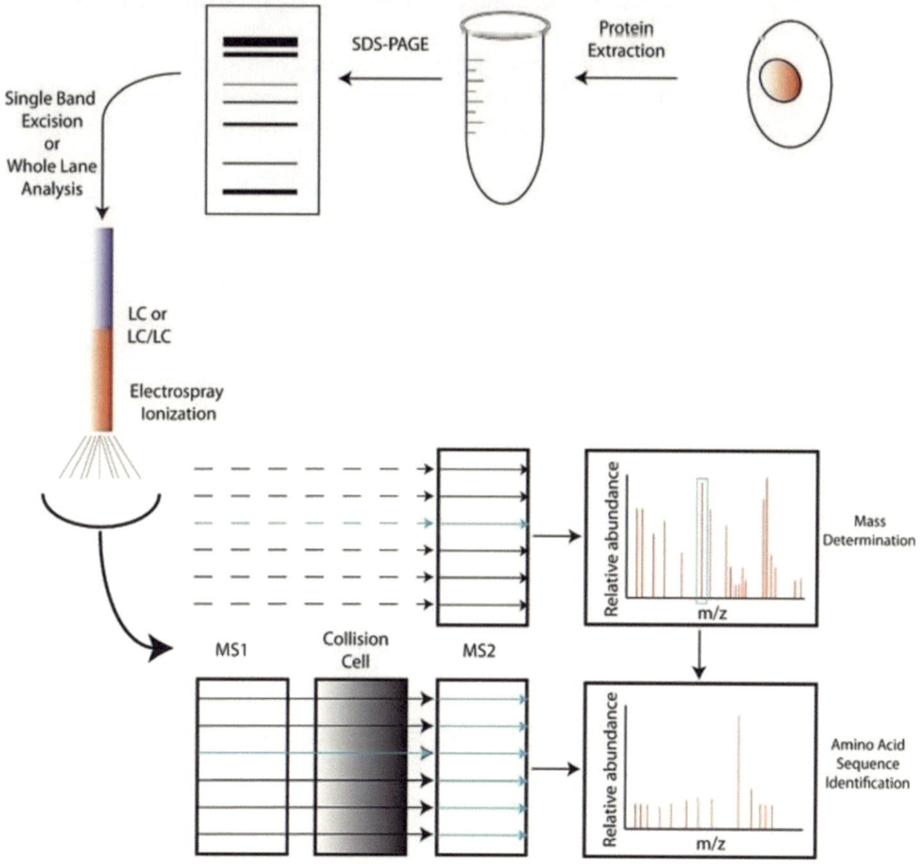

Figure 3.
Representation depiction of LC-MS procedure by Wang et al. [23] is licensed under CC BY 3.0.

times the amount that can be accommodated by the vacuum system of the mass spectrometer [5, 7, 10].

The second point is that in LC, sample volatility is not an issue. Therefore, for the coupling between the techniques to become useful, the mass spectrometer must use an ionization technique that produces ions in the gaseous phase without necessarily requiring thermal vaporization of the sample [10]. During the last few years, several ionization techniques have been developed for MS providing optimal ionization of non-volatile molecules, such as those already mentioned [4–10, 20–22].

Given the limitations, one could question the role of the LC-MS technique (and its derivations such as HPLC, UHPLC, HILIC, MS/MS, etc.). Is it an attractive instrumentation that generates a multitude of data that no one can reproduce? Or is it a useful technology that will serve as a basis for routine and quality testing? Will it be (or perhaps already is) established as the main chemical analysis tool?

It is clear that LC-MS/MS is evolving very quickly (also based on remarkable technological advances and large investments by manufacturers). The fact is that mass spectrometry is taking the place of other technologies and it looks like it will eventually replace them in the different types of laboratories and fields of application of analytical chemistry. **Figure 3** shows the schematic of LC-MS procedure.

3. Sample preparation and other analytical issues in LC-MS analysis

Despite the development of automated methods, much of sample preparation still depends on manual human effort, so that this step of the analytical process is the most error-prone, and the most time and resource-consuming in modern chemical analysis [10, 24].

Sample preparation is considered by many scientists to be the work of unskilled personnel, and, as a result, there is a tendency to apply rather outdated sample preparation methods or protocols and liquid-liquid extraction schemes that use large volumes of solvents [10, 24]. As a result, the adoption of new techniques becomes urgently needed.

The more extensive the sample preparation in a complex matrix, the more targeted and less comprehensive the analysis. But, of course, minimal sample preparation is always indicated, for representativeness of the compounds of interest, reducing matrix interference or co-elution, among other important requirements for LC-MS analyses [7, 24, 25].

For analysis of solid samples, for example, the majority only need to be solubilized to obtain a liquid solution to be submitted to the analytical system [10, 24]. When we talk about other types of samples, extraction procedures are employed, such as liquid-liquid extraction, solid phase extraction, solid phase microextraction, supercritical fluid extraction, and specific extraction using polymers [7, 10, 24, 25].

Overall, this depends a lot on the characteristics of that sample. For urine, normally centrifugation and dilution are sufficient [20, 22]. However, for blood-derived samples (e.g., plasma and serum), proteins must be precipitated using organic solvents to avoid compromising the integrity of the LC-MS system [11, 19, 20].

The application of the mass spectrometer does not mean that we will be able to analyze all the relevant compounds in a sample, since not all of them are ionizable [8, 24]. A portion of these compounds are always disadvantaged and discriminated against, even for comprehensive ionization strategies. The signal strength of the compounds varies depending on the matrix, sample type, and system performance [22, 24].

Derivatization is a chemical modification of the analyte(s) in the sample leading to the stabilization of the molecule(s) in the chromatographic analysis, reducing separation problems during the analytical run [5, 22]. When HPLC-MS analyses are used, great care must be taken and very well understood about the molecules of interest. Derivatization can lead to characteristic fragmentation applicable, for example, to distinguish positional isomers, when we talk about LC-MS [22].

Another common tool in LC-MS studies is the use of standards labeled with stable isotopes of the compounds of interest [22, 26]. However, not all classes of these compounds are commercially available in this form. Combining a labeled and unlabeled analyte leads to twin peaks with a known mass difference, which is useful for studying fragmentation pathways in tandem mass spectra (MS/MS) [5, 10, 26].

Many LC methods without MS detection often rely on the use of non-volatile buffers such as phosphate buffers and other inorganic additives. These non-volatile compounds cannot be recommended for HPLC-MS coupling due to possible contamination of the ion source and also due to ionic suppression (matrix effect) [12, 16, 22].

Inorganic acids in the mobile phase must be replaced by organic acids (formic or acetic acid) with typical concentrations below 0.1%, but sometimes higher [5, 8, 11]. For an alkaline environment, ammonium hydroxide in a similar concentration range can be used with alkaline pH extended stability columns for LC [10, 11]. Typical volatile salts in HPLC-MS/MS are ammonium acetate or formate, with a common concentration range of 2–10 mmol/L [5, 8, 10, 11].

4. Applications of the LC-MS/MS technique

That the LC-MS/MS technique has high resolving power, accuracy, and detection of analytes in complex samples, or at low concentrations, has already been discussed. In this part of the chapter, will be presented some outstanding applications in the area and its advances.

The first application cited for analysis by mass spectrometry is in the study of xenobiotic chemical compounds (those that can be harmful to human and animal health as a whole). These compounds may be present in drugs, industrial chemicals, pollutants, pesticides, pyrolytic products from cooked foods, secondary plant metabolites (polyphenols, flavonoids, glycosides, terpenes, steroids, alkaloids, and antibiotics) and toxins from bacteria, fungi, plants, and animals [1, 8, 9, 11, 17, 20, 22, 24–26].

As it is known, due to the complexity and diversity of metabolite properties, a single analytical technique cannot be applied to detect all thousands of metabolites in a biological sample [11, 17, 22, 24, 26]. However, the detection of these xenobiotic compounds by masses must be preceded by separation techniques such as HPLC, in the HPLC-MS/MS junction [1, 8, 9, 11, 17, 22, 24]. Thus, it is possible to perform the structural discrimination of positional isomers and stereoisomers based on different retention times [17, 20, 22, 24].

The metabolomics community realized that the application of hyphenated techniques is an appropriate strategy to analyze most of the detected metabolites that are in different ranges of polarity and molecular weight [22, 24–26].

The GC-MS technique is widely used for metabolomic studies and provides efficient and reproducible analyses [1, 8, 9, 11, 17, 20]. However, non-volatile compounds that do not derivatize and thermolabile compounds will not be observed in the

GC-MS analysis, which attracts even more interest for the addition of HPLC-MS/MS as an analytical tool, mainly for the analysis of human blood plasma [9, 11, 22, 24].

UHPLC-MS/MS-based metabolomics has also been applied to profile urinary metabolites for processes related to toxicity and pathogenesis, with tests in rats [11, 17, 22, 26]. It was realized that the introduction of UHPLC coupled to high-resolution mass spectrometers provides high analytical accuracy for the discovery and identification of target metabolites or biomarkers. However, the implementation of the technique still requires substantial experience and know-how, being limited to a very small number of routine clinical laboratories.

To further extend the coverage of the metabolome, APCI ionization has shown the potential to reveal some additional metabolites, which exhibit high ionization potential in this mode compared to ESI [1, 8, 9, 11, 17, 20]. Most global metabolite profiling studies are currently performed on TOF-MS machines due to the sensitivity, rapid data acquisition, and high mass accuracy they offer [8, 9, 11, 17, 20, 22].

The ability of the mass spectrometer to separate molecules based on their mass-to-charge ratio (m/z) has led many researchers to conclude that results can be obtained more quickly using direct introduction [i.e., direct infusion/direct insertion] and that the separation of molecules on the basis of their m/z is sufficient, especially when using very high-resolution mass spectrometers [1, 8, 20, 24–27]. As an example, the application of UHPLC-MS/MS resulted in the quantitative determination of >140 polar metabolites in a single injection of wine or fruit extracts [27].

When we look at environmental samples, with regard to the presence of pollutants in aquatic and terrestrial ecosystems, some compounds can spread through the environment contaminating living organisms, which may affect the health of these ecosystems [8, 9, 24, 25, 27]. In short, the HPLC-MS/MS technique has shown the best results in most common cases such as monitoring of drugs, cosmetics, and related preservatives [8, 9, 27].

These are just some of the cited applications for such a rich and important technique in the analysis of chemical compounds in different matrices and areas of study.

5. The future of analytics power

Currently, as applications based on miniaturized instrumental analyses cover several areas of research (toxicology, medicine, food safety and quality, environmental surveillance, among others), work aimed at the development of new technologies that make separations by nano-LC commonplace is also encouraged [12, 13, 16].

In summary, these automated systems work as follows: first, the sample is loaded onto the miniaturized extraction column in order to clean the sample, extract, and pre-concentrate the compounds of interest. After that, a valve is turned to the elution position, and the compounds are transferred to the miniaturized analytical column, promoting chromatographic separation and detection [12–15].

New approaches such as microchips are emerging as highly miniaturized column types, suitable for field or real-time analyses. This increase in the search for new stationary phases and analytical column technologies is helping miniaturized LC to become popular and to cover several fields of application of analytical chemistry [15, 16].

The main objectives of miniaturization are to increase portability, reliability, analysis speed, and jointly reduce costs, as well as simplify the chromatographic step making it more understandable for inexperienced operators [12–15]. The fact of being able to gather most of the LC components in a micro-sized planar structure, combined with the reduced number of connections in the fluidic system, can mitigate problems of chromatographic peak broadening [11–14].

As is widely known, both capillary-LC and nano-LC have notable advantages when compared to standard HPLC, including increased sensitivity (decreasing LOD and LOQ values), as well as a decrease in the need for stationary phase, in solvent consumption and waste generation. Furthermore, recent investigations have focused on the development of miniaturized sample preparation methods to reduce the required sample volume [3, 12–16]. **Figure 4** shows a miniaturized system of nano-LC.

Automation using an online coupling approach, integrating sample preparation and chromatographic separation, has been shown to avoid multi-step procedures and time-consuming analyses. For these reasons, combining the advantages of LC and sample preparation on a miniaturized scale emerges as an attractive approach to perform direct extraction, separation, detection, and quantification of target analytes [13, 14].

Despite the known benefits of chip-based LC, there are also several challenges to be overcome, such as simplifying procedures for making chips in the laboratory; process reproducibility; accuracy and precision in the injection as well as in the pumping system; in addition to coupling to sample detection or pre-treatment steps [3, 11–14].

It should be noted that such a reduction in the analytical work scale requires a high sensitivity detector suitable for working at low flow rates and lower concentration levels. And in this regard, the mass spectrometer (MS) is in practice the detector responsible for meeting the current sensitivity requirements, thus becoming an ideal detector for miniaturized LC [3, 11–16].

Currently, several different nano-ESI systems are commercially available, designed to improve ion emissions by creating constant flows from the sample through the miniaturized LC to the ion source [12, 13, 15, 16]. However, even with this successful coupling between the two techniques, the analysis of complex samples still requires more precise and selective results, which has driven the application of tandem mass spectrometry (MS/MS) at this juncture.

This "tandem" configuration results in higher levels of specificity [3, 11, 12]. Furthermore, reducing the flow rate to the order of microliters or nanoliters per minute results in the formation of sub-micrometer droplets. With this, there is an improvement in the ionization of the analytes and in the signal-to-noise response [15, 16].

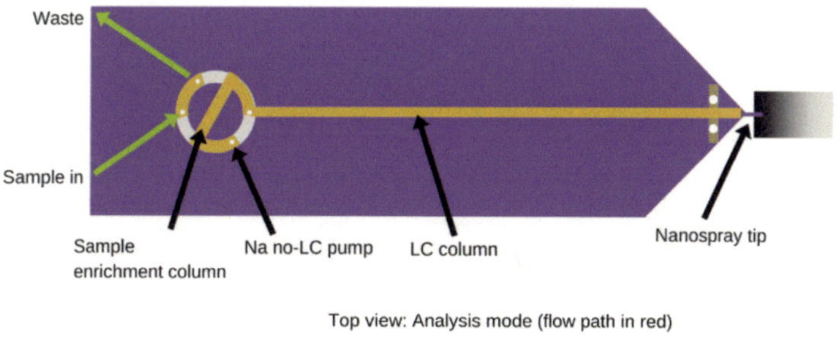

Figure 4.
Schematic of how a nano-LC system. Note that the system is made up of the main parts that a conventional HPLC has, such as sampler, column, and flow system (pump).

Despite the known advantages of mass spectrometry coupled to miniaturized LC, the complexity of their physical communication still represents an obstacle to the advancement of these systems. This obstacle occurs mainly because the LC works with a higher pressure in the liquid phase, while the MS requires a vacuum condition for proper operation [3, 11, 12, 15].

6. Conclusions

With everything that has been presented in this chapter, we can consider that even if we cannot obtain a configuration between LC, with regard to ion sources and mass analyzers, fully comprehensive for the different fields of application of chemical analysis, the hyphenated technique LC-MS/MS still represents the most promising way forward. On the one hand, selectivity of the analyte and reproducibility of separation by chromatography (HPLC, UHPLC, HILIC, etc.) participate, and on the other hand, we can generate embedded chemical information, confirming the identity of the analytes based on their molecular mass and specific ion fragments, by through tandem mass spectrometry. On miniaturization, the development and improvement of miniaturized or portable LC-MS systems can represent a significant achievement for the "in the field" analysis of forensic, environmental, food, medical, military and public interest exposures.

Conflict of interest

The authors declare no conflict of interest.

Author details

Kelly da Silva Bezerra
Universidade Federal de Goiás, Goiás, Brazil

*Address all correspondence to: kelly_bezerraa@hotmail.com

References

[1] Kukusamude C, Burakham R, Chailapakul O, Srijaranai S. High performance liquid chromatography for the simultaneous analysis of penicillin residues in beef and milk using ion-paired extraction and binary water-acetonitrile mixture. Talanta. 2012;**92**:39-44. DOI: 10.1016/j.talanta.2012.01.020

[2] Pirok BWJ, Stoll DR, Schoenmakers PJ. Recent developments in twodimensional liquid chromatography: Fundamental improvements for practical applications. Analytical Chemistry. 2019;**91**:240-263. DOI: 10.1021/acs.analchem.8b04841

[3] Miro M, Hansen EH. Miniaturization of environmental chemical assays in flowing systems: The lab-on-a-valve approach vis-a-vis lab-on-a-chip microfluidic devices. Analytica Chimica Acta. 2007;**600**:46-57. DOI: 10.1016/j.aca.2007.02.035

[4] Hollender J, Schymanski EL, Singer HP, Ferguson PL. Nontarget screening with high resolution mass spectrometry in the environment: Ready to go? Environmental Science & Technology. 2017;**51**:11505-11512. DOI: 10.1021/acs.est.7b02184

[5] Shackleton C. Clinical steroid mass spectrometry: A 45-year history culminating in HPLC-MS/MS becoming an essential tool for patient diagnosis. The Journal of Steroid Biochemistry and Molecular Biology. 2010;**121**:481-490. DOI: 10.1016/j.jsbmb.2010.02.017

[6] Morlock GE. High-performance thinlayer chromatography combined with effect directed assays and high-resolution mass spectrometry as an emerging hyphenated technology: A tutorial review. Analytica Chimica Acta. 2021;**1180**:1-14. DOI: 10.1016/j.aca.2021.338644

[7] Ardrey B, Wiley JS. Liquid Chromatography-Mass Spectrometry: An Introdution. New Jersey, USA: Wiley; 2003. 296 p. ISBN: 0-47149801-7

[8] Steiner D, Sulyok M, Malachov A, Mueller A, Krska R. Realizing the simultaneous liquid chromatography-tandem mass spectrometry based quantification of >1200 biotoxins, pesticides and veterinary drugs in complex feed. Journal of Chromatography. A. 2020;**1629**:461-502. DOI: 10.1016/j.chroma.2020.461502

[9] Hernández F, Sancho JV, Ibáñez M, Abad E, Portolés T, Mattioli L. Current use of high-resolution mass spectrometry in the environmental sciences. Analytical and Bioanalytical Chemistry. 2012;**403**:1251-1264. DOI: 10.1007/s00216-012-5844-7

[10] Menger F, Gago-Ferrero P, Wiberg K, Ahrens L. Wide-scope screening of polar contaminants of concern in water: A critical review of liquid chromatography-high resolution mass spectrometry-based strategies. Trends in Environmental Analytical Chemistry. 2020;**28**:1-11. DOI: 10.1016/j.teac.2020.e00102

[11] Maurer HH. What is the future of (ultra) high performance liquid chromatography coupled to low and high resolution mass spectrometry for toxicological drug screening? Journal of Chromatography A. 2012;**1292**:19-24. DOI: 10.1016/j.chroma.2012.08.069

[12] Sikanen TM, Kiiski I, Ollikainen E. Microfluidic analysis techniques for safety assessment of pharmaceutical

Nano- and Microsystems. In: Peltonen L, editor. Characterization of Pharmaceutical Nano- and Microsystems. New Jersey, USA: Wiley-VCH Verlag; 2021. pp. 97-136

[13] Harendarcíková L, Baron D, Sebestová A, Rozsypal J, Petr J. True lab-in-asyringe technology for bioassays. Talanta. 2017;**174**:285-288. DOI: 10.1016/j.talanta.2017.06.017

[14] Wink K, Mahler L, Beulig JR, Piendl SK, Roth M, Belder D. An integrated chip-mass spectrometry and epifluorescence approach for online monitoring of bioactive metabolites from incubated Actinobacteria in picoliter droplets. Analytical and Bioanalytical Chemistry. 2018;**410**:7679-7687. DOI: 10.1007/s00216-018-1383-1

[15] Maciel EVS, de Toffoli AL, Sobieski E, Nazario CED, Lanças FM. Miniaturized liquid chromatography focusing on analytical columns and mass spectrometry: A review. Analytica Chimica Acta. 2020;**1103**:11-31. DOI: 10.1016/j.aca.2019.12.064

[16] Lynch KB, Chen A, Liu S. Miniaturized high-performance liquid chromatography instrumentation. Talanta. 2018;**177**:94-103. DOI: 10.1016/j.talanta.2017.09.016

[17] Zhang Y, Zhang Y, Wang G, Chen WS, He P, Wang Q. Simultaneous separation of polar and non-polar mixtures by capillary HPLC based on an ostadecylsilane and taurine derivatized silica continuously packed column. Talanta. 2016;**161**:762-768. DOI: 10.1016/j.talanta.2016.09.022

[18] Zhang M, Gong L, Liu G, Kang J. Preparation of a monolithic column with a mixed-mode stationary phase of reversed-phase/hydrophilic interaction for capillary liquid chromatography. Journal of Separation Science. 2019;**42**:662-669. DOI: 10.1002/jssc.201800826

[19] Rainville PD, Theodoridis G, Plumb RS, Wilson ID. Advances in liquid chromatography coupled to mass spectrometry for metabolic phenotyping. Trends in Analytical Chemistry. 2014;**61**:181-191. DOI: 10.1016/j.trac.2014.06.005

[20] Holčapek M, Kolářová L, Nobilis M. High-performance liquid chromatography–tandem mass spectrometry in the identification and determination of phase I and phase II drug metabolites. Analytical and Bioanalytical Chemistry. 2008;**391**:59-78. DOI: 10.1007/s00216-008-1962-7

[21] Cole RB. Electrospray and MALDI Mass Spectrometry: Fundamentals, Instrumentation, Practicalities, and Biological Applications. New Jersey, USA: John Wiley & Sons; 2010. DOI: 10.1002/9780470588901

[22] Theodoridis GA, Gika HG, Want EJ, Wilson ID. Liquid chromatography–mass spectrometry based global metabolite profiling: A review. Analytica Chimica Acta. 2012;**711**:7-16. DOI: 10.1016/j.aca.2011.09.042

[23] Wang J, Trowbridge JJ, Rao S, Orkin SH. Proteomic studies of stem cells. In: StemBook, editor. The Stem Cell Research Community. StemBook; 2008. DOI: 10.3824/stembook.1.4.1. Available from: http://www.stembook.org

[24] Gago-Ferrero P, Gros PM, Ahrens L, Wiberg K. Impact of on-site, small and large scale wastewater treatment facilities on levels and fate of pharmaceuticals, personal care products, artificial sweeteners, pesticides, and perfluoroalkyl substances in recipient waters. Science of the Total

Environment. 2017;**601-602**:1289-1297. DOI: 10.1016/j.scitotenv.2017.05.258

[25] Kennedy DG, Mccracken RJ, Cannavan A, Hewitt SA. Use of liquid chromatography-mas spectrometry in the analysis of residues of antibiotics in meat and a milk. Journal of Chromatography. 1998;**812**:77-98. DOI: 10.1016/s0021-9673(98)00048-x

[26] Zhao S, Li L. Chemical isotope Labeling LC-MS for metabolomics. Advances in Experimental Medicine and Biology. 2021;**1280**:1-18. DOI: 10.1007/978-3-030-51652-9_1

[27] Vidal JLM, Liébanas FJA, Rodríguez MJG, Frenich AG, Moreno JLF. Validation of a gas chromatography-triple quadrupole mass spectrometry based method for the quantification of pesticides in food commodities. Rapid Communications in Mass Spectrometry. 2006;**20**:365-375. DOI: 10.1002/rcm.2315

Section 2

Applications

Chapter 5

Green Strategies toward Eco-Friendly HPLC Methods in Pharma Analysis

Natalija Nakov, Jelena Acevska, Katerina Brezovska, Zoran Kavrakovski and Aneta Dimitrovska

Abstract

The global need for changing the processes in order to meet the green analytical chemistry (GAC) criteria is a great challenge for the pharmaceutical industry. High-performance liquid chromatography (HPLC), as one of the most frequently used techniques in various stages in the pharmaceutical industry, generates huge amounts of organic toxic waste. Therefore, the implementation of the GAC principles in pharma analysis is highly required. Although the number of published papers concerning green chromatography approaches is constantly increasing, the use of eco-friendly HPLC methods in the pharma industry has not been widely implemented. The reasons for this mainly include the need for adaptation of the conventional HPLC instruments, lack of time, lack of experience, or uncertainty of the analysts regarding fulfillment of the method criteria. In this chapter, an overview of green strategies that can be easily applied to conventional instruments for liquid chromatography (LC) in developing eco-friendly HPLC methods in pharma analysis is given. The aim is to emphasize that the green method development in pharma analysis can be easily accomplished and to encourage the analytical community in the pharmaceutical industry not only to develop but also to transfer the already established conventional HPLC methods into green ones.

Keywords: green analytical chemistry, reversed-phase liquid chromatography, pharmaceutical analysis, green solvents, micellar liquid chromatography, per aqueous liquid chromatography, ethanol, propylene carbonate, glycerol, surfactants

1. Introduction

The concept of "green chemistry" (GC) refers to the design of chemical processes and products that enable elimination (or reduction) of the use or creation of substances that are harmful to humans and the environment. The GC concept is based on 12 principles set out by Paul Anastas and John Warner in 1998 [1]. An extensive overview of these principles could be found on the American Chemical Society (ACS) webpage [2]. The term "green analytical chemistry" (GAC) was introduced in 2001 [3]. The GAC refers to the development of new, effective analytical methodologies that will enable minimization and/or elimination of hazardous chemicals and chemical waste while enabling faster and more energy-efficient analysis. Therefore, the above-mentioned 12 principles of green chemistry were modified to formulate the main features determining the green character of the analytical chemistry and place it in the function of GAC. The GAC principles could be summarized in four main topics: (1) elimination (or reduction) of the consumption of reagents from analytical procedures; (2) minimization of energy consumption; (3) proper management of analytical waste; and (4) increased safety for the operator [2–4]. Scientists, researchers, and analysts across the world recognized the need for implementation of the GAC principles in analytical methods, which resulted in constant increase of the number of published papers in different fields of GAC (fundamentals, spectroscopy, electrochemistry, and separation methods) [5].

The pharma industry is not immune to the global need of changing the processes to meet GAC criteria. The study conducted in 2019 by researchers from McMaster University in Canada showed that the pharmaceutical industry globally emitted a higher amount of greenhouse gases compared to the automotive production sector [6]. Pharma analysis is a fundamental part of the pharmaceutical industry, whereas high-performance liquid chromatography (HPLC) is one of the most commonly used techniques. HPLC finds application in various stages of the lifecycle of medicines, starting with pharmacokinetic, pharmacodynamics, and bioequivalent studies; quality control of active pharmaceutical ingredients (APIs) and excipients; control of the manufacturing process; quality control of the finished product; stability studies; and so on.

During the development and validation of the chromatographic methods, focus is placed on the chromatographic parameters such as accuracy, precision, and robustness, as well as the analysis runtime. However, other aspects regarding the impact of the chromatographic method on the analyst's safety and the environment are still insufficiently taken into account.

The following text summarizes the green strategies (along with their strengths and drawbacks) that are available for analysts in the development of eco-friendly chromatographic methods that can be immediately applied to conventional HPLC instruments. This summary will contribute toward the easy acceptance of the need for transformation of the already-established conventional HPLC methods in the pharma industry into "green" chromatographic methods.

2. Tools for evaluation of the greenness of chromatographic methods

The implementation of the GAP principles in the processes of green chromatographic method development, as well as the transfer of the conventional HPLC methods into eco-friendly solutions, is an essential part of a development strategy of the pharma industry. Alongside the green method development, it is also equally important to consider the available approaches aimed at evaluating the greenness of the developed methods. The analyst should take into consideration the greenness of the proposed method and compare it with the existing ones. This kind of evaluation provides quantitative data regarding how well each of the different segments of the method conforms to the GAC principles. In this way, the analyst can have additional information for further method improvement in terms of critical method parameters.

At the beginning of the implementation of the GAC principles in chromatographic methods, the analysts did not address this issue enough. Over the years, more and more emphasis has been placed on the greenness assessment of the published methods in the field of pharma analysis. Several tools are used for evaluation of the different aspects of the methods and their compliance with the GAC principles, providing general assessment (e.g., National Environment Methods Index) or quantitative estimation (e.g., analytical eco-scale). In this section, a brief description of the tools that are most commonly used for evaluation of the greenness of the chromatographic method is presented.

The National Environment Methods Index (NEMI) is one of the oldest available tools for the assessment of the greenness of methods [7]. This approach evaluates the method in four different fields: persistent, bio-accumulative and toxic (PBT); hazard; corrosive; and waste. The greenness profile of the method is presented as a simple pictogram, divided into four sections. The color of each section can be green (if the term meets the particular criteria) or black (if the term does not meet the green criteria). In this manner, a general assessment is obtained, which is very easily visualized. However, the NEMI index does not take into consideration energy consumption, as one of the GAC principles. Another disadvantage of this assessment tool is that it cannot be considered a qualitative tool, and in cases where non-typical chemicals are used, the process of preparation of the symbols is more time-consuming.

In 2012, a quantitative tool named analytical eco-scale index had been presented [8]. According to this index, a perfectly green method is assigned with 100 points (theoretically ideal green method). A lower number of points for a given method indicate that the method has a larger deviation from the GAC principles and lower greenness. The analytical eco-scale points can be calculated by subtracting the number of penalty points from 100. The penalty points are calculated on various grounds. For example, the number of penalty points given for the reagents depends on the type and the volume of reagents used for the procedure. The penalty points assigned for energy consumption are determined by the type of instrumentation used. For example, titration is a technique that consumes the least amount of energy, while liquid chromatography-mass spectrometry (LC-MS) and gas chromatography-mass spectrometry (GC-MS) are the most energy-consuming techniques, leading to higher penalty points. After summarizing the total number of penalty points for the different segments of the method, the analytical eco-scale points are obtained. A score above 75 points represents an excellent green method; a score ranging from 75 to 50 indicates an acceptable green method, while a score below 50 designates inadequate green analysis. The analytical eco-scale index, as a quantitative tool, facilitates the process of comparison of the greenness of the methods. In addition, the calculation of penalty points is simple, and it allows analysts to assess which part of the procedure has the major contribution toward nonconformity with GAC principles. The analytical eco-scale approach has been widely accepted by scientists.

The green analytical procedure index (GAPI) is a tool that enables evaluation of the greenness of the whole analytical procedure, starting from the sample collection to the final determination [9]. This tool takes into account the advantages of the previously discussed tools, such as NEMI and eco-scale index. The GAPI tool visually is presented in terms of five pentagrams, which represent the different segments of the analytical procedure, and they are further subdivided into 15 different segments. These pentagrams can be colored green, yellow, or red, depending on their environmental impact. The first pentagram refers to the sample, and it is subdivided into four parts, representing the collection (1), preservation (2), transportation (3), and storage (4) of the sample. The second pentagram is positioned in the middle, and it refers to the type of the method (5), whether the method is direct or indirect. The third pentagram refers to the sample preparation, and it includes the scale of extraction (6), used solvents (7), and additional treatments (8). The fourth pentagram refers to the reagents, and it is subdivided into three parts: amount (9), health hazard (10), and safety hazard (11). The last pentagram refers to the instrumentation, and it contains five parts: energy (12), occupation hazard (13), waste (14), and waste treatment (15). The GAPI index is very comprehensive; thus, it can be considered as a qualitative and a quantitative tool for the method of greenness assessment.

The analytical method greenness score (AMGS) metric is an online, freely accessible calculator, which can be found on the ACS webpage [10]. This web-based calculator is only applicable for LC methods (normal-phase LC, ultra-pressure LC, LC-MS, and preparative LC-MS) and supercritical fluid chromatography (SFC) methods such as preparative SFC and SFC-MS. The authors [11] introduced this tool for the assessment of the environmental, health, and safety impact (EHS) score of the solvents, the solvent energy demand, and the instrument energy consumption. The AMGS score is automatically calculated when the users enter several pieces of information about the method, that is: the type of technique used, number of compounds of interest, number of injections, flow rate, rut time of the analysis, solvents used for the mobile phase, type and volume of the sample solvent, and so on. A lower AMGS score implies a greener method. In addition, the users get quantitative information about the impact of the three main categories (instrument energy, solvent energy, and solvent safety) on the score. These three main categories can be colored green, yellow, or red, depending on their contribution to the AMGS value. The yellow and red colors indicate the area where the method can be improved. The current limitations of the AMGS calculator are that this tool is not applicable in cases where more than three components are used in the mobile phase and in cases where techniques different from LC and SFC are used. However, taking into account that LC has the widest application in the pharma industry and that the process of gathering information about the greenness score of the method is very simple, AMGS calculator is considered as a very useful tool.

In 2020, the analytical greenness (AGREE) metrics approach and software were published [12]. This metric system takes into account the 12 principles of GAC, so the method is assessed for each principle separately. The 12 GAC principles are converted into a numerical value (score) ranging from 0 to 1. The overall AGREE score is a product of the individual results for each GAC principle. In addition to the quantitative result, the AGREE score is presented in the form of a pictogram, enclosing the final result in the middle of the pictogram. The performance of the method regarding the 12 variables is presented as 12 segments in the outer part of the pictogram. The segments are in a color scale (green-yellow-red), and the width of each corresponding segment in the pictogram reflects the contribution of the variables to the AGREE score. The AGREE approach is a very comprehensive tool for the greenness method assessment because it takes into account each of the 12 principles, but at the same time, it is a very simple tool, allowing easy interpretation of results. The same group of authors published a tutorial for the use of the AGREE metrics for sample preparation [13].

3. Green solvents as a mobile phase

The reverse phase high performance liquid chromatography (RP-HPLC) as a technique requires a large number of organic solvents as eluents in the mobile phase.

It is well known that methanol (MeOH) and acetonitrile (ACN) are the most consumed organic solvents in the RP-HPLC mobile phase. These solvents are hazardous, and both have acute and chronic toxic effects. Methanol is a toxic alcohol that can cause retina damage and serious acidosis [14, 15]. Acetonitrile toxicity is manifested through inhalation of vapors or contact with skin and eyes. Acetonitrile in vivo is metabolized to cyanide, leading to cytotoxic anoxia [16]. Another issue with the ACN is its price on the market. This reagent is a byproduct of acrylonitrile, and the decreased demand for acrylonitrile in 2008 led to reduced production and a shortage of ACN [17]. Therefore, the price of acetonitrile on the market increased many times after the ACN crisis. It becomes evident that the use of these solvents in the mobile phase raises issues related to the health of the analysts as well as brings increased expenses for the analytical labs. Additionally, the negative ecological impact of these solvents cannot be neglected because of the huge amount of chemical waste. The chemical waste generated per year by HPLC instruments worldwide is approximately 34 million liters [18].

Considering these data, it becomes understandable that the removal of these toxic solvents from the RP-HPLC mobile phases and their replacement with greener alternatives have a key role in the development of eco-friendly HPLC methods. In addition, solvents used for the sample preparation steps for one HPLC method should also be taken into consideration. It is well known that the polarity of the solvents used for sample preparation should match the polarity of the mobile phase to obtain symmetrical chromatographic peaks. Therefore, the composition of the solvent for the sample preparation process should correspond to the composition of the mobile phase. Taking into account the abovementioned, it becomes evident that the waste generated from the whole HPLC procedure should be considered during the ecological assessment of the HPLC method.

Several organic solvents such as ethanol (EtOH), acetone, ethyl acetate, 2-propanol, glycerol, and propylene carbonate (PC) are used as green alternatives for conventional organic eluents in the mobile phase. However, not all of the mentioned green organic alternatives have the same advantages and allow easy method transfer from conventional to eco-friendly HPLC methods. Namely, the main disadvantage of acetone and ethyl acetate is that they have very high UV cutoffs (330 and 260 nm, respectively), so their use becomes incompatible with UV detector. Another issue is their high viscosity, leading to higher column back pressure [19–21].

In this chapter, green organic solvents that are easily affordable and that produce good chromatographic performance, such as ethanol, propylene carbonate, and glycerol, will be discussed.

3.1 Ethanol

Several review papers have highlighted the advantages of the use of ethanol as a green solvent in RP-HPLC mobile phases [21–24]. Although ethanol, according to Snyder's classification of organic solvents [25], belongs in the same group as methanol, this solvent can be used as a replacement for acetonitrile too. Ethanol has several advantages over these two solvents.

Regarding human health effects, ethanol has lower vapor pressure, so the toxic effects of inhalation of ethanol vapors are reduced. The toxic effect of ethanol is more related to long-term ingestion and not to its use as a reagent. Considering the environmental impact, ethanol is biodegradable and has less negative environmental impact compared to acetonitrile and methanol [21].

From an analytical point of view, ethanol has higher eluotropic strength, so lower quantities are needed to achieve comparable retention time to acetonitrile and methanol. The UV cutoff is acceptable, and it's around 210 nm [25]. The main drawback is the higher viscosity of the ethanol-water mobile phase, leading to higher column backpressure. Our experience shows that this column backpressure can be easily overcome with the use of higher column temperatures of 35 or 40°C or with lower flow rates. Considering the GAP principles for reducing reagent consumption, the lower flow rate will not be an issue but an added value to the green method.

From an economic point of view, the market price of ethanol is lower than those of acetonitrile and methanol, so the method cost is lower. In addition, since ethanol is less toxic, the waste disposal cost is lower. This goes in favor of the overall reduction of expenses in the pharmaceutical analysis.

All these features have contributed to ethanol being the most preferred green alternative for acetonitrile and methanol. There are numerous examples of the application of ethanol-based green mobile phase in RP-HPLC in pharma analysis. Yabré et al. gave an extensive review of the pharmaceutical applications regarding the use of ethanol-based mobile phase [21].

This chapter gives an overview of certain publications, from 2018 to date, concerning the use of ethanol-based HPLC mobile phase in pharma analysis. In 2022, a comparative study of two HPLC methods for the determination of four antipsychotics (quetiapine fumarate, aripiprazole, asenapine maleate, and chlorpromazine hydrochloride) was published [26]. The authors developed one green and one conventional RP-HPLC method for the determination of these antipsychotics in bulk and pharmaceutical formulations (Quitapex®, Asenapine®, Aripiprazole®, and Neurazine® tablets). The mobile phase for the green method consisted of ethanol and 20 mM sodium dihydrogen phosphate in a ratio of 35:65, *v/v* (pH 5.0), and the separation was achieved in isocratic mode within 11 minutes. For the conventional HPLC method, acetonitrile was used instead of ethanol, and the separation of the studied antipsychotics was achieved in gradient mode in 15 minutes. The flow rate for both methods was 1 mL/min. Both methods were validated following International Conference on Harmonization (ICH) Q2R1 Guideline [27]. It was observed that the green HPLC method showed better limits of detection (LOD) and limits of quantification (LOQ) for the analyzed antipsychotics. In addition, the evaluation of the obtained results using Student's t-test and F-ratio showed that there was no statistically significant difference between the accuracy and the precision of the green method and the conventional HPLC method. The greenness of the proposed methods was evaluated using the GAPI index. The green HPLC method for the determination of the four antipsychotics had nine green and six yellow pentagrams, while the conventional method had six green, six yellow, and three red pentagrams. This research demonstrates that the ethanol-based LC mobile phase not only gives benefits from an ecological point of view but also leads to better method performances such as shorter runtime and better LOD and LOQ.

Another group of authors applied the design of experiment (DoE) approach to optimize a green HPLC method for the determination of atorvastatin in tablets and in the presence of its degradation products. The optimized mobile phase consisted of ethanol and 0.5% aqueous acetic acid (*v/v*) in a ratio of 57.5–42.5, with a flow rate of 0.91 mL/min and a column temperature of 40°C. The method was validated, and the statistical evaluation of the assay results showed that there was no significant difference between the proposed and the reference methods. The greenness of the method was evaluated using four different approaches: GAPI, AMGS, analytical eco-scale, and AGREE software-based tool. The applied approaches suggested that the method was more environmentally friendly compared to other published HPLC methods for the determination of atorvastatin [28].

The numerous benefits of the analytical quality by design (AQbD) approach contributed to the more extensive use of this methodology for the green method development. The Central Composite Design (CCD) model has been applied for the optimization of the chromatographic conditions for the stability-indicating RP-HPLC method for the determination of escitalopram and etizolam in tablets [29]. The ethanol-potassium dihydrogen phosphate buffer (60:40 v/v, respectively) mobile phase allows the separation of the APIs from their degradation products obtained under stress conditions (acidic and alkaline hydrolysis, oxidation, photodegradation, and dry heat). The greenness of the proposed method was confirmed by four different assessment methods: NEMI, GAPI, AMGS, and AGREE.

In another study, QbD methodology was also used for method development for the simultaneous determination of impurities of artesunate and amodiaquine as APIs [30]. The method development was complex mostly due to the differences in the polarity of the impurities of the analyzed APIs. Another challenge was the low UV absorption maximum (210 nm) for artesunate and its impurities, leading to poor detection with ethanol-based mobile phase driven by the UV cutoff of ethanol (210 nm). The QbD approach allowed the development of a simple and robust HPLC method using an ethanol-based mobile phase. The separation of nine impurities, including acidic, basic, and structurally related compounds, was achieved using ethanol and 10 mM acetic acid in gradient mode. The analytical eco-score of the proposed method was 95, confirming that the method complied with the GAP principles.

Another research that was published recently also demonstrated the applicability of the ethanol-water mobile phase for the simultaneous determination of three different APIs in tablets [31]. In this study, famotidine, paracetamol, thiocolchicoside, and caffeine (as internal standard) were separated using ethanol and 50 mM sodium dihydrogen phosphate (pH 4.6) in gradient mode. The method was validated following ICH guidelines [27]. The robustness testing showed that the method was robust. The proposed method was compared in terms of method performance and greenness with other published methods for the determination of the studied compounds. The evaluation showed that the proposed method was eco-friendly and had the same or better LOD/LOQ compared with previously published methods. Another paper [32] presents the use of an ethanol-water mobile phase (containing 0.05% triethanolamine at pH 4.5, in a ratio of 90:10, respectively) for the determination of two different moxifloxacin combinations (moxifloxacin/dexamethasone and moxifloxacin/prednisolone). The optimized and validated green method was successfully applied for the determination of the studied APIs in an eyedrop solution, ophthalmic and otic solution, as well as ophthalmic suspension. As for the previously mentioned papers, the authors of this paper also made a comparison between the results obtained from the proposed green method and the reference methods using the Student t-test and the variance ratio F-test. The results showed that there was no statistically significant difference between the proposed and the reference methods. The authors of this paper also evaluated the greenness of the proposed method using the three most-often-used approaches: NEMI, GAPI index, and analytical eco-scale index. In addition, there is a detailed tabular view for assigning and counting the eco-scale penalty points for the proposed and previously reported methods for the studied compounds. The eco-scale score was found to be 94, indicating excellent agreement with the GAC principles.

Ethanol can be used as a green alternative to acetonitrile even for chiral separations. An example of the use of ethanol-based chiral mobile phase is the stability-indicating RP-HPLC method for simultaneous determination of timolol and latanoprost in eye drops [33]. The mobile phase consists of β-cyclodextrin, sodium octane sulphonate, and ethanol in gradient mode. The optimized mobile phase enabled the separation of the two APIs in the presence of latanoprost C3-epimer, latanoprost acid, and timolol-related compounds C and E and from the degradation products obtained under stress conditions. Considering that the organic solvent influences the inclusion of the β-cyclodextrin, the successful replacement of acetonitrile with ethanol opens an opportunity for the use of this green solvent in chiral separations. The detection wavelength was set at 210 nm, showing that the detection near the UV cutoff of ethanol was applicable. The column temperature was 35°C, while the flow rate was 1.2 mL/min. The issue with the higher column backpressure was solved with the use of monolithic columns. The monolithic columns are packed with a single piece of porous material (called "silica rod"), which fills the column, so there are no interparticle voids. This type of design enables higher total porosity, higher separation efficiency, and lower pressure drop. A major advantage of the monolithic columns is that they can work under high flow rates (up to 10 mL/min) without generation of high column backpressure [34]. The author highlighted the applicability of the use of monolithic columns as an approach for dealing with the high column pressure generated from the ethanol-based mobile phase. The method validation showed that the method was robust. The green analysis assessment was made using the GAPI pictogram.

Table 1 summarizes some of the recently published articles regarding the use of ethanol as a green chromatographic eluent for pharma analysis.

3.2 Propylene carbonate

Propylene carbonate belongs to the group of cyclic carbonated solvents, and it is derived from carbon dioxide. This solvent, as a green aprotic solvent with high polarity, can be used as a replacement for the toxic aprotic polar solvents such as acetonitrile, dimethyl formamide, or dimethyl sulfoxide [42, 43]. PC was used for the first time as an organic modifier in the LC mobile phase in 2011 [44]. Since then, there are only a few articles regarding the use of this green solvent in the LC mobile phase [45–47]. The investigations showed that the problem with the limited miscibility of PC with water could be solved by using a mixture of PC with methanol or ethanol. Considering the GAC concept, ethanol should be considered as a third solvent in the PC/water mobile phase. The mixture of PC/EtOH/water has acceptable viscosity. In addition, the PC has an acceptable UV cutoff of 210 nm.

The most recent research that demonstrates the applicability of PC as a replacement for ACN has focused on the separation of a mixture of 39 pharmaceutical compounds in two-dimensional liquid chromatography (2D LC) [48]. The authors revealed that the effective peak distribution and peak capacity obtained with PC as an organic modifier were comparable to those obtained using ACN and MeOH. Despite the significantly reduced runtime (32 min for methods with PC *vs* 52 min for methods with ACN and MeOH), an improvement in separation selectivity was achieved [48]. This research confirmed the role of propylene carbonate as an effective substitute for acetonitrile, providing greater separation power even in 2D LC. The results from this research open a possibility for wider applications of this solvent in the future.

API	Medicinal product	Mobile phase	Column	Reference
Antipsychotics	Tablets	EtOH: 20 mM NaH_2PO_4 pH 5.0 = 35:65	Thermo C18 250 × 4.6 mm, 5 μm	[26]
Atorvastatin calcium	Bulk, Tablets	0.5% Acetic acid: EtOH = 42.5:57.5	Zorbax Eclipse Plus C18 150 × 4.6 mm, 5 μm	[28]
Escitalopram and etizolam	Tablets	EtOH: KH_2PO_4 pH 2.5 = 60:40	C18	[29]
Artesunate and amodiaquine impurities	Tablets	Aqueous solution: ethanol, gradient elution	XBridge BEH C18 150 × 3 mm 5 μm	[30]
Famotidine, paracetamol and thiocolchicoside	Tablets	50 mM NaH_2PO_4 pH 4.6 and EtOH, gradient elution	C8 150 × 4.6 mm, 5 μm	[31]
Moxifloxacin/ dexamethasone and moxifloxacin/prednisone	Eye drop solution, Ophthalmic suspension, Ophthalmic and otic solution	EtOH: water = 90:10, containing 0.05% triethanolamine, pH 4.5	Hypersil C8 250 × 4.6 mm, 5 μm	[32]
Timolol and latanoprost	Eye drop solution	(β-cyclodextrin + sodium octane sulphonate): EtOH = 40:60, gradient elution	Kinetex XB-C18 150 × 4.6 mm	[33]
Ketoconazole and beclomethasone	Bulk, Cream formulation	EtOH: 0.1 M KH_2PO_4 pH 2.5 = 33:67	ODS 250 × 4.6 mm, 5 μm	[35]
Pyridoxine HCl and doxylamine succinate	Tablets	EtOH: 0.01 M phosphate buffer pH 5.0 = 10:90	Xterra C18 100 × 4.6 mm, 5 μm	[36]
Tafluprost	Bulk, Ophthalmic formulation	EtOH: 0.01 M phosphate buffer pH 4.5 = 60:40	Hyperclone C18 150 × 4.6 mm, 5 μm	[37]
Lamivudine, zidovudine and nevirapine	Tablets	EtOH and 0.1 M ammonium acetate pH 4.5, gradient mode	C18 250 × 3.0 mm, 5 μm	[38]
3,4-methylenedioxymetham phetamine (MDMA)	Tablets	H_2O with 0.1% formic acid, pH 5.0: EtOH = 85:15	ACE-5 Phenyl	[39]
Rosuvastatin	Tablets	EtOH: MeOH: EtAc = 6:3:1	Nucleodor C8 150 × 4.6 mm, 5 μm	[40]
Secnidazole	Tablets	(H_2O + 0.7% acetic acid): EtOH = 78:22	Luna CN 250 × 4.6 mm, 5 μm	[41]

Table 1.
Examples of HPLC methods using ethanol-water mobile phase in pharma analysis.

3.3 Glycerol

Glycerol is an organic solvent that is the most recently included in the development of green chromatographic methods. In 2021, glycerol was used for the separation of four antiviral medicines under reversed-phase chromatographic conditions [49]. The same group of authors used this solvent as a mobile phase modifier for the determination of ascorbic acid and glutathione in tablets [50].

Several features make this solvent suitable for the development of eco-friendly LC methods. First of all, glycerol is a nonvolatile safe solvent, so the risk to the analyst's health due to vapor inhalation is reduced. The advantage regarding lab safety is that under normal storage conditions, glycerol has low flammability and high stability. The benefit in terms of environmental protection is that it is biodegradable, and it is available from renewable, cheap sources [51, 52]. According to the CHEM21 selection guide of classical and less classical solvents, the health, safety, and environmental scores of ethanol are 3, 4, and 3, respectively, whereas for glycerol, these scores are 1, 1, and 7, respectively [43]. From an analytical point of view, glycerol has an advantage over ethanol and PC in terms of UV cutoff and miscibility with water. Namely, glycerol has a UV cutoff of 207 nm, which is slightly lower compared to ethanol, and it is completely miscible with water in comparison to PC, which has limited miscibility.

The higher viscosity of glycerol could be seen as a drawback from one side, but on the other hand, it could bring a potential benefit. Due to the higher viscosity, it is recommended for glycerol to be premixed with the water phase, followed by sonication. This step will reduce the pump load of the LC system and will facilitate the mixing process [49, 50]. In addition, the issue with higher viscosity could be overcome with the use of higher column temperature, because this parameter has a significant impact on the column back pressure. For instance, the increase of the column temperature of 10°C led to a 20% decrease in column back pressure. The research results regarding the influence of the glycerol viscosity on the terms from the Van Deemter equation showed that higher viscosity of the glycerol-based mobile phase led to a decrease of the eddy diffusion (A term) and a decrease of the longitudinal diffusion (B term) [49]. This could be the reason for the good chromatographic performance in terms of peak symmetry, column efficiency, and resolution. Therefore, in this case, the higher viscosity of the glycerol-based mobile phase could be seen as a potential benefit. Regarding the elution strength, glycerol is in the middle between water as a weak eluent and acetonitrile and methanol as strong eluents in RP-HPLC. This property allows analysts to use glycerol for better adjustments of the elution strength of the mobile phase and to obtain better selectivity when needed.

4. Pure water as a mobile phase

Everyone would agree that there is no greener option for making the chromatographic analysis eco-friendlier than choosing pure water as LC mobile phase. This idea is present in the analytical community for more than 30 years [53], but its popularity has grown in the last decade.

Two approaches enable the use of pure water as a LC mobile phase. The first approach is to apply elevated temperatures along with stationary phases that are stable under these conditions. It should be considered that under the term "elevated temperature", a temperature above 100°C or water heated below its critical point conditions (374°C and 218 atm) should be considered [54]. In the literature studies, this type of chromatography could be found under the name of superheated water chromatography (SHWC) or subcritical water chromatography (SCWC) [54, 55]. Subcritical water has different properties compared to water under ambient temperature conditions. Namely, the temperature increase leads to a decrease in the dielectric constant, viscosity, and surface tension of the water. The result is a decrease in the water polarity, so in the chromatographic system, it begins to behave more like the organic solvents. Therefore, the nonpolar compounds can be eluted under higher water temperatures, whereas lower temperatures are needed for the elution of polar compounds [54, 56]. Considering the limited temperature stability of the classical RP columns, the use of pure water under subcritical conditions requires the application of thermally stable stationary phases that could withstand the temperatures above 200°C. Some examples of stationary phases that could be applied for SCWC are polymer-based (such as polystyrene and divinylbenzene), zirconia-based, or carbon-based stationary phases [57].

The application of SHWC is related to the need for the adaptation of conventional HPLC instruments. One of the most crucial instrumental modifications is the requirement of special column ovens that can heat the column very fast, as well as detectors other than UV/Vis (e.g., amperometric detector, flame ionization detector, etc.) [54]. The need for instrument modification could be one of the reasons that this technique has not had wider application in pharma analysis yet.

The second approach is based on the use of pure water under ambient temperature conditions (below 60°C) [54, 58]. In such cases, there is no need for adaptation of the conventional HPLC instruments. In addition, the thermal stability of the stationary phases is not an issue anymore. The only requirement is to use columns that are stable under high water content surrounding. If silica-based stationary phases are used, then the term per aqueous liquid chromatography (PALC) should be used [59]. PALC should be distinguished from hydrophilic interaction liquid chromatography (HILIC) because although in PALC and HILIC the same type of stationary phases is used (silica-based columns), in HILIC, the water content in the mobile phase is very low, whereas in PALC, water is the dominant eluent in the mobile phase [53, 55, 57]. In PALC, the surface of the silica stationary phase becomes nonpolar in conditions where the water content in the mobile phase is very high. As explained in the literature [59], this effect is due to the higher participation of siloxane groups.

A variant of PALC is the water-only reversed-phase liquid chromatography (WRP-LC) [53]. The difference between PALC and WRP-LC is that the second one uses polar-embedded or polar-endcapped stationary phases instead of silica-based columns.

In cases like this, where the choice of the mobile phase composition is very limited (pure water or high water content), the selectivity is mostly controlled by the type of stationary phase. In the recent years, several different polar-embedded and polar-endcapped stationary phases have appeared on the market. The differences between the polar-embedded and polar-endcapped stationary phases are presented in comprehensive way [57]. The available polar-embedded and polar-endcapped stationary phases have different solvation properties, resulting in differences in selectivity. The importance of proper selection among the various types of polar-embedded stationary phases (Amino-P-C18, Diol-Ester C10, Diol-Ester C18, and Diol-Ester Phenyl) was presented by the separation of polar compounds (nucleic bases, nucleosides, and purine alkaloids) using pure water as a mobile phase [54]. Separation of four amino acids (hydroxyproline, proline, glycine, and alanine) using pure water as a mobile phase was achieved using a mixed-mode polar embedded column [60]. The stationary phase used for this application has long alkyl chains and ion-exchange functional groups attached at the terminal end of this chain, thus providing the reverse phase and ion-exchange nature of this column [60]. This kind of stationary phase could increase the possibilities for the separation of components with different polarities, thus making WRP-LC as an approach for green chromatographic method development more attractive.

5. Surfactants as a mobile phase

The amphiphilic molecules have two different ends (hydrophilic, polar and hydrophobic, nonpolar) with an opposite affinity for the dispersion medium. These molecules can reduce surface tension, so they are called surface-active substances or surfactants. These molecules, when present in low concentrations in the dispersion medium, exist as individual molecules of a sub-colloidal size (monomers). By increasing the concentration of amphiphilic molecules, their aggregation occurs, and micelles are formed. The concentration of monomers necessary to form micelles is called the critical micellar concentration (CMC). Theoretically, micelles are formed at a certain concentration of monomers, but in practice, this happens in a narrow concentration range [61].

Surfactants have diverse applications in different fields of analytical chemistry [62]. Further text includes a brief overview of the application of surfactants as green eluents in the mobile phase. The chromatographic technique that uses surfactants as a mobile phase is called micellar liquid chromatography (MCL). This technique was introduced in 1980 [63], but it was "forgotten" during the past years. However, the increased awareness of the GAC principles contributed to the return of MLC in the focus of the analysts. Surfactants used in MLC are nontoxic, biodegradable, and have low environmental bioconcentration factors [64]; therefore, their use as eluents in the RP-HPLC mobile phase is considered as an effective strategy for greening the HPLC methods.

The MLC is a type of reversed-phase chromatography; hence, conventional nonpolar stationary phases (C18, C8, etc.) are used. The mobile phase consists of an aqueous solution of surfactants in a concentration above their CMC. As this mobile phase travels through the column, it forms two different constituents: the micelles (also called micellar pseudophase) and the surfactant monomers that are present in the aqueous environment. In addition, surfactants modify the surface of the stationary phase: the hydrophobic end of the surfactants binds to the nonpolar stationary phase, whereas the hydrophilic end is directed toward the mobile phase. The surfactants' orientation on the surface of the stationary phase forms a kind of an open micelle structure. Therefore, the polarity and the charge of the surface of the stationary phase depend on the nature of the surfactant used (anionic, cationic, or nonionic). The analytes are separated based on their different partitioning between the modified stationary phase, the pseudophase, and the bulk solvent. The unique characteristics of the surfactants, as well as their position and effects on the stationary phase, enable the existence of different kinds of retention mechanisms (hydrophobic, ionic, and steric), allowing separation of analytes with different polarities [65, 66].

The most commonly used surfactants as eluents in MCL are the anionic sodium dodecyl sulfate (SDS), the cationic cetyl trimethyl ammonium bromide (CTAB), and the nonionic polyoxyethylene-23-lauryl ether (Brij-35). These surfactants have low CMC (e.g., 8.2 mM for SDS and 0.09 mM for Brij-35) [65], providing acceptable viscosity of the mobile phase. The most important chromatographic parameters that should be considered during the method development are the column temperature, the type and concentration of the surfactant used, and the type of co-eluents, if needed. The working column temperature should be above the Kraft point (the temperature at which the solubility of the amphiphilic molecules equals the CMC). Considering that the Kraft point for the commonly used surfactants is on ambient temperature (e.g., for SDS, the Kraft point is on 15°C) [65] and that the viscosity of the mobile phase is temperature dependent, the chromatographic separation is usually performed between 25 and 60°C. The increased temperature contributes to faster mass transfer kinetics, which often improves the separation efficiency [67]. Generally, the increase in the column temperature leads to a shorter retention time of the analytes [68, 69]. However, in some cases, it was observed that higher column temperature contributed to the increased value of the tailing factor [69].

The surfactant concentration usually has an inverse influence on the retention time, due to the stronger association of the analytes with the micelles [70]. An exception to this dependence is the chromatographic behavior of phenolic compounds, where the concentration of Brij-35 has a positive correlation with the retention time. This could be explained by a hydrogen bond formation between the phenolic compounds and the hydroxyl groups of Brij-35; thus, increase in the retention time is observed [71, 72].

Co-eluents in the surfactant-based mobile phase are usually needed to decrease the retention of the nonpolar compounds and to improve the mass transfer to the stationary phase. Short-chain alcohols such as methanol, ethanol, n-propanol, n-butanol, n-pentanol, or acetonitrile in low concentrations (3–15%) are frequently used to enhance the elution strength of the surfactant-based mobile phase [73]. These alcohols act by reducing the ability of the monomers to form micelles; thus, the ability of the analyte to bind the micelles is decreased. This effect is more pronounced when long-chain alcohols are used [62, 66]. This finding was also confirmed during the determination of two ternary mixtures (phenylephrine hydrochloride, ibuprofen, and chlorpheniramine maleate as mixture 1 and pseudoephedrine hydrochloride, ibuprofen, and chlorpheniramine maleate as mixture 2) utilized for cold treatment [74]. In this case, use of methanol and acetonitrile in the mobile phase provided longer retention times of the analytes, but asymmetrical peaks were obtained, which was not the case when n-propanol was used.

In recent years, in order to fulfill the GAC criteria, the organic alcohols have been replaced with the addition of more polar surfactants (e.g., Brij-35) in the mobile phases. The combination of two types of surfactants in the LC mobile phase (usually SDS and Brij-35) modifies the MLC. This modified MLC is called mixed-mode micellar liquid chromatography (mixed MLC). The mixed MCL allows the separation of neutral and charged analytes. Brij-35, as a more polar surfactant compared to SDS, decreases the polarity of the stationary phase, leading to reduced retention time of the polar compounds. The SDS, as an anionic surfactant, provides a negative charge to the surface of the stationary phase; thus, polar compounds with a positive charge bind more strongly, and consequently, their retention time is longer. Brij-35, as a nonpolar surfactant, reduces this negative charge of the modified stationary phase; thus, the column keeps its neutral character [72]. The mixed micellar mobile phase was used to separate 10 commonly used antihypertensive medicines (hydrochlorothiazide, chlorthalidone, atenolol, losartan, amiloride, valsartan, spironolactone, olmesartan, bisoprolol, and irbesartan) using C18 core-shell column [75]. An interesting finding is that the increase in the concentration of the surfactants did not have the same effect on the retention of all target compounds. The expected behavior is that the increase in the concentration of Brij-35 will decrease the retention time of the analytes (higher elution power of the mobile phase). However, the authors noticed that longer retention time of hydrochlorothiazide was observed with the increase of concentration of Brij-35 from 0.01 M to 0.04 M. On the other hand, valsartan first showed a decrease in the retention time, and afterward, the retention time increased. Similar behavior but in the opposite direction has been observed for spironolactone. For this compound, the increase of the concentration of Brij-35 from 0.01 M to 0.03 M led to a longer retention time, and when the concentration of Brij-35 was 0.04 M, the retention time of this compound decreased. This finding shows that in mixed MLC, it is very hard to predict the retention behavior of the analytes; thus, the method development becomes a more challenging process. Considering that various factors affect the retention behavior of the analytes and that their effect could not be easily predicted, the use of the DoE approach during the method development process is highly recommended.

The DoE approach was used for the development of a robust mixed MLC method for the determination of five antidiabetic medicines (metformin, glipizide, glimepiride, pioglitazone, and repaglinide) using Symmetry C18 column [64]. The use of CCD design allowed fast method optimization, and at the same time, the requirements for the method parameters were fulfilled. The separation of the medicines was accomplished in 10 minutes. Another method uses the same experimental design for the optimization of a mixed MLC method for the determination of six medicines (paracetamol, guaifenesin, pseudoephedrine, ibuprofen, chlorpheniramine, and dextromethorphan) for the treatment of common cold [72]. The response surface methodology proved to be successful in this case as well, allowing the separation of six analytes with adequate resolution in a short analysis time. The application of the DoE approach during the mixed MLC method development facilitates the optimization of the critical chromatographic parameters such as concentration of SDS and Brij-35 and allows timely perception of their interaction. Polysorbates are another type of nonionic surfactant that can be used for MLC, including polysorbate 20 (Tween 20) or polysorbate 40 (Tween 40) [68, 73]. The comparison of the influence of Tween 20 and Tween 40 as eluents in the MLC mobile phase on the retention of hydroxycinnamic acid compounds showed that shorter retention times and better peak shapes are observed using Tween 20 [68].

The gradient elution in MLC is possible by changing the concentration of either the surfactant or the organic solvent (if present). In both cases, the time needed for re-equilibration of the column is shorter, and analysis time is faster when compared to the gradient mode using conventional mobile phases. The faster re-equilibration time is due to the constant amount of surfactant being absorbed on the surface of the stationary phase. During the gradient elution, the increased monomer concentration leads to greater micelle formation, so the concentration of the monomer does not change, and it is kept around the CMC value [66, 73]. The gradient elution mode in MLC is not very suitable for nonpolar hydrophobic compounds, because they are more strongly retained. This could be overcome by the inclusion of an organic solvent or more polar surfactant as a co-eluent or with the use of more polar reversed-phase stationary phases.

It is very important that the analysts who are starting to use MLC on their HPLC systems for the first time provide information about the column conditioning, column care, and special considerations regarding surfactant mobile phase flushing. Detailed instructions for the care of the chromatographic system in MLC can be found in literature studies [65, 76]. In brief, the reversed-phase column should be first flushed with 100% water (at least 30 column volumes) to remove any present organic solvent. Afterward, the surfactant-based mobile phase can be introduced in the system. The column equilibration with this kind of mobile phase takes a longer time. Another consideration that should be taken into account is that the mobile phase should be continuously flushed through the column. If the flushing stops, there is a great possibility of precipitation of the surfactant and consequently crystal formation around the seals of the pump or in the column. To prevent system blockage or column clogging, the pumps should work continuously overnight. To minimize solvent consumption, the flow rate should be set to a minimum (e.g., 0.1 mL/min) [65, 76]. After finishing the analysis, the HPLC system as well as the chromatographic column should be gradually washed up to 100% water to remove the surfactant. As a further step, the column should be restored according to the instructions given by the manufacturer for the column conditioning.

6. Conclusion

As presented in this chapter, the green strategies needed for the development of eco-friendly HPLC methods in pharma analysis could be easily accomplished and applied to conventional LC instruments.

Ethanol, as the most extensively used green organic solvent, provides better method performances (shorter runtime, better LOD, better LOQ, etc.) compared to conventional organic solvents (acetonitrile and methanol). Many of the published papers presented that there was no statistically significant difference between the results (regarding the method accuracy and precision) obtained with the ethanol-based methods and the reference methods based on the conventional LC mobile phase. Propylene carbonate is still not an extensively used green alternative, but recent publications have shown that analysts could take into account this solvent as an effective substitute for acetonitrile, even for 2D LC. Glycerol, as the most recently used green alternative for conventional organic solvents, has an elution strength that is in between the elution strength of water and methanol/acetonitrile. Even though the number of published papers concerning the use of glycerol is still scarce, analysts should consider this solvent as an option for green method development.

If we speak about the greenest solvent for the LC mobile phase, then pure water is the solution. The increased availability of modern materials that are used for the production of polar-embedded and polar-endcapped stationary phases resulted in increased applicability of PALC and WPR-LC as valuable techniques for eco-friendly HPLC method development.

Many papers demonstrate that MLC methods, based on surfactant mobile phase, fully meet the GAC criteria, allowing separation of compounds with different polarity. The main challenge in this type of green chromatography is the method development process. The use of the DoE approach facilitates the optimization of the critical chromatographic parameters and leads to faster and easily accomplished MLC method development. Regardless of the type of the mobile phase (ethanol, pure water, or surfactants), it is recommended to use the DoE methodology for the method optimization process, because this approach reduces the number of experiments, and it follows the GAP principles.

Different tools (such as NEMI, GAPI, Eco-scale index, AMGS, and AGREE) used for the evaluation of the greenness of the HPLC methods unambiguously demonstrate that eco-friendly methods have advantages over the conventional methods in terms of ecological impact, operator's safety, and energy consumption. Along with ecological and economic benefits, the eco-friendly methods provide better method performances, being an additional motivation for implementation of the GAC concept in the R&D departments and quality control labs in the pharma industry. The implementation of the green strategies in the pharma analysis will provide benefits for the analysts (healthier working environment), the pharma industry itself (lower method cost and lower waste disposal costs), and the whole microcommunity (reduced negative environmental impact).

Acknowledgements

The authors acknowledge the Faculty of Pharmacy, Ss. Cyril and Methodius University in Skopje, Republic of North Macedonia, for supporting the publication process of this chapter.

Conflict of interest

The authors declare no conflict of interest.

Author details

Natalija Nakov*, Jelena Acevska, Katerina Brezovska, Zoran Kavrakovski and Aneta Dimitrovska
Institute of Applied Chemistry and Pharmaceutical Analysis, Faculty of Pharmacy, Ss. Cyril and Methodius University in Skopje, Republic of North Macedonia

*Address all correspondence to: natalijan@ff.ukim.edu.mk; natalija.nakov82@gmail.com

References

[1] Anastas P, Eghbali N. Green chemistry: Principles and practice. Chemical Society Reviews. 2010;**39**:301-312. DOI: 10.1039/b918763b

[2] American Chemical Society. Available from: https://www.acs.org/content/acs/en/greenchemistry/principles/12-principles-of-green-chemistry.html. [Accessed: September 30, 2022]

[3] Namiesnik J. Green analytical chemistry—Some remarks. Journal of Separation Science. 2001;**24**:151-153. DOI: 10.1002/1615-9314(20010201)24:2

[4] Gałuszka A, Migaszewski Z, Namiesnik J. The 12 principles of green analytical chemistry and the significance mnemonic of green analytical practices. Trends in Analytical Chemistry. 2013;**50**:78-84. DOI: 10.1016/j.trac.2013.04.010

[5] Guardia M, Garrigues S. Chapter 1: Past, present, and future of green analytical chemistry. In: Garrigues S, Guardia M, editors. Challenges in Green Analytical Chemistry. 2nd ed. From Book Series: Green Chemistry Series. Cambridge, UK: Royal Society of Chemistry; 2020. pp. 1-18. DOI: 10.1039/9781788016148-00001

[6] Nawrat A. Turning pharma green: An eco-wish list for the industry. Analyst. 2020

[7] Keith L, Gron L, Young J. Green analytical methodologies. Chemistry Review. 2007;**107**:2695-2708. DOI: 10.1021/cr068359e

[8] Galuszka A, Konieczka P, Migaszewski Z, Namiesnik J. Analytical eco-scale for assessing the greenness of analytical procedures. TrAC Trends in Analytical Chemistry. 2012;**37**:61-72. DOI: 10.1016/j.trac.2012.03.013

[9] Plotka-Wasylka J. A new tool for the evaluation of the analytical procedure: Green analytical procedure index. Talanta. 2018;**181**:204-209. DOI: 10.1016/j.talanta.2018.01.013

[10] ACG Green Chemistry Institute, AMGS Spreadsheet Calculator. Available from: https://www.acsgcipr.org/amgs/. [Accessed: October 31, 2022]

[11] Hicks MB, Farrell W, Aurigemma C, Lehmann L, Weisel L, Nadeau K, et al. Making the move towards modernized greener separations: Introduction of the analytical method greenness score (AMGS) calculator. Green Chemistry. 2019;**21**:1816-1826. DOI: 10.1039/C8GC03875A

[12] Pena-Pereira F, Wojnowski W, Tobiszewski M. AGREE—Analytical GREEnness metric approach and software. Analytical Chemistry. 2020;**92**:10076-10082. DOI: 10.1021/acs.analchem.0c01887

[13] Pena-Pereira F, Tobiszewski M, Wojnowski W, Psillakis E. A tutorial on AGREE prep an analytical greenness metric for sample preparation. Advances in Sample Preparation. 2022;**3**:100025. DOI: 10.1016/j.sampre.2022.100025

[14] Medinsky MA, Dorman DC. Recent developments in methanol toxicity. Toxicology Letters Volumes. 1995;**82-83**:707-711. DOI: 10.1016/0378-4274(95)03515-X

[15] Souza FGT, Nogueira VVE, Maynart LI, Oliveira RL, Mendonça TCS, Oliveira PD. Optic neuropathy toxic after methanol inhalation. Brazilian Journal of Ophthalmology. 2018;**77**:47-49. DOI: 10.5935/0034-7280.20180010

[16] Joshi DR, Adhikari N. An overview of common organic solvents and their toxicity. Journal of Pharmaceutical Research International. 2019;**28**:1-18. DOI: 10.9734/jpri/2019/v28i330203

[17] Majors R. The continuing acetonitrile shortage: How to combat it or live with it. LCGC North America. 2009;**27**:458-471

[18] Tobiszewski M. Metrics for green analytical chemistry. Analytical Methods. 2016;**8**:2993-2999. DOI: 10.1039/C6AY00478D

[19] Funari C, Carneiro R, Khandagale M, Cavalheiro A, Hilder E. Acetone as a greener alternative to acetonitrile in liquid chromatographic fingerprinting. Journal of Separation Science. 2015;**38**:1458-1465. DOI: 10.1002/jssc.201401324

[20] Micale F, Albu F, Lorgulescu EE, Medvedovici A, Tache F. Ethyl lactate as a greener alternative to acetonitrile in RPLC: A realistic appraisal. Journal of Chromatographic Science. 2015;**53**:1701-1707. DOI: 10.1093/chromsci/bmv077

[21] Yabre M, Farey L, Some I, Gaudin K. Greening reverse-phase liquid chromatography methods using alternative solvents for pharmaceutical analysis. Molecules. 2018;**23**:1065. DOI: 10.3390/molecules23051065

[22] Olives A, Gonzalez-Ruiz V, Martin M. Sustainable and eco-friendly alternatives for liquid chromatography analysis. ACS Sustainable Chemistry & Engineering. 2017;**5**:5618-5634. DOI: 10.1021/acssuschemeng.7b01012

[23] Tobiszewski M, Namiesnik J. Greener organic solvents in analytical chemistry. Current Opinion in Green and Sustainable Chemistry. 2017;**5**:1-4. DOI: 10.1016/j.cogsc.2017.03.002

[24] Shaaban H. New insights into liquid chromatography for more eco-friendly analysis of pharmaceutical. Analytical and Bioanalytical Chemistry. 2016;**408**:6929-6944. DOI: 10.1007/s00216-016-9726-2

[25] Snyder LR, Kirkland J, Dolan JW. Introduction to Modern Liquid Chromatography. 3rd ed. Hoboken, New Jersey, USA: John Wiley & Sons; 2009. ISBN 978-0-470-50818-3

[26] Hameed E, El-Naby Z, Gindy A, Zaitone S, Alshaman R, Saraya R, et al. Two new HPLC methods, assessed by GAPI, for simultaneous determination of four antipsychotics in pharmaceutical formulations: A comparative study. Separations. 2022;**9**:220. DOI: 10.3390/separations9080220

[27] ICH Harmonised Tripartite Guideline. Validation of Analytical Procedures: Text and Methodology Q2(R1). ICH Expert Working Group. Geneva, Switzerland: International Conference on Harmonisation of Techical Requirements for Registration of Pharmaceuticals for Human Use; 2005

[28] Kokilambigai K, Lakshmi K. Analytical quality by design assisted RP-HPLC method for quantifying atorvastatin with green analytical chemistry perspective. Journal of Chromatography Open. 2022;**2**:100052. DOI: 10.1016/j.jcoa.2022.100052

[29] Perumal D, Krishnan M, Lakshmi KS. Eco-friendly based stability-indicating RP-HPLC technique for the determination of escitalopram and etizolam by employing QbD approach. Green chemistry Letters and Reviews. 2022;**15**:671-682. DOI: 10.1080/17518253.2022.2127334

[30] Yabré M, Ferey L, Somé T, Sivadier G, Gaudin K. Development of a green HPLC method for the analysis of

artesunate and amodiaquine impurities using quality by design. Journal of Pharmaceutical and Biomedical Analysis. 2020;**190**:113507. DOI: 10.1016/j.jpba.2020.113507

[31] Dogan A, Eylem C, Akduman N. Application of green methodology to pharmaceutical analysis using ecofriendly ethanol-water mobile phases. Microchemical Journal. 2020;**157**:104895. DOI: 10.1016/j.microc.2020.104895

[32] Ibrahim F, Elmansi H, Fathy M. Green RP-HPLC method for simultaneous determination of moxifloxacin combinations: Investigation of the greenness for the proposed method. Microchemical Journal. 2019;**148**:151-161. DOI: 10.1016/J.MICROC.2019.04.074

[33] Ibrahim A, Saleh H, Elhenawee M. Assessment and validation of green stability indicating RP-HPLC method for simultaneous determination of timolol and latanoprost in pharmaceutical dosage forms using eco-friendly chiral mobile phase. Microchemical Journal. 2019;**148**:21-26. DOI: 10.1016/j.microc.2019.04.059

[34] Díaz-Bao M, Barreiro R, Miranda J, Cepeda A, Regal P. Recent advances and uses of monolithic columns for the analysis of residues and contaminants in food. Chromatography. 2015;**2**:79-95. DOI: 10.3390/chromatography2010079

[35] Kannaiah KP, Sugumaran A. Environmental benign AQbD based estimation of ketoconazole and beclomethasone by RP-HPLC and multi-analytical UV spectrophotometric method. Microchemical Journal. 2022;**172**:106968. DOI: 10.1016/j.microc.2021.106968

[36] El-Hadi H, Eltanany B, Zaazaa H, Eissa M. HPLC-DAD approach for determination of pyridoxine HCl and doxylamine succinate in pure and pharmaceutical dosage forms: A green stability-indicating assay method. Microchemical Journal. 2022;**172**:106982. DOI: 10.1016/j.microc.2021.106982

[37] Abd-AlGhafar W, Aly F, Sheribah Z, Saad S. Green, validated HPLC method coupled with fluorescence detection for the analysis of tafluprost in its pure form and ophthalmic formulation: Application to aqueous humor and content uniformity testing. Journal of Chromatographic Science. 2022:bmac061. DOI: 10.1093/chromsci/bmac061 [Epub ahead of print]

[38] Vieira-Sellaï L, Quintana M, Diop Q, Mercier O, Tarrit S, Raimi N, et al. Green HPLC quantification method of lamivudine, zidovudine and nevirapine with identification of related substances in tablets. Green Chemistry Letters and Reviews. 2022;**15**:695-704. DOI: 10.1080/17518253.2022.2129463

[39] Duarte L, Ferreira B, Silva G, Ipólito A, de Oliveira M. Validated green phenyl reverse-phase LC method using ethanol to determine MDMA in seized ecstasy tablets. Journal of Liquid Chromatography & Related Technologies. 2020;**43**:761-769. DOI: 10.1080/10826076.2020.1811725

[40] Nazrul H, Faiyaz S, Fars A, Doaa HA, Abbas IM. Development and validation of a green RP-HPLC method for the analysis of rosuvastatin: A step towards making liquid chromatography environmentally benign. Green Processing and Synthesis. 2018;**7**:160-169. DOI: 10.1515/gps-2017-0023

[41] Lima J, Kogawa AC, Salgado HRN. Green analytical method for quantification of secnidazole in tablets by HPLC-UV. Drug Analytical Research. 2018;**02**:20-26

[42] Byrne F, Jin S, Paggiola G, Petchey T, Clark J, Farmer T, et al. Tools and techniques for solvent selection: Green solvent selection guides. Sustain Chemical process. 2016;**4**:7. DOI: 10.1186/s40508-016-0051-z

[43] Prat D, Wells A, Hayler J, Sneddon H, McElroy C, Abou-Shehada S, et al. CHEM21 selection guide of classical and less classical-solvents. Green Chemistry. 2016;**18**:288-296. DOI: 10.1039/C5GC01008J

[44] Suvarna B, Namboodiry V, Pratibha V, Soni M, Ashok B. Prospective use of propylene carbonate as a mobile phase component in RP-HPLC. Chemistry. 2011. Corpus ID: 99784898

[45] Varsha N, Suvarna B, Pratibha V, Soni M, Ashok B. Replacement of acetonitrile by mixtures of propylene carbonate and methanol as organic modifier in mobile phases for RPLC separation mechanism: Application to the assay of alprazolam and sertraline in combined pharmaceutical formulations. Journal of Liquid Chromatography & Related Technologies. 2012;**35**:2643-2654. DOI: 10.1080/10826076.2011.637273

[46] Varsha N, Pratibha V, Soni M, Ashok B, Suvarna B. Estimation of paracetamol and lornoxicam by isocratic, gradient, and elevated temperature HPLC using propylene carbonate. Journal of Liquid Chromatography & Related Technologies. 2014;**37**:1094-1103. DOI: 10.1080/10826076.2013.765464

[47] Tache F, Udrescu S, Albu F, Micale M, Medvedovici A. Greening pharmaceutical applications of liquid chromatography through using propylene carbonate-ethanol mixtures instead of acetonitrile as organic modifier in the mobile phases. Journal of Pharmaceutical and Biomedical Analysis. 2013;**75**:230-238. DOI: 10.1016/j.jpba.2012.11.045

[48] Aly A, Górecki T, Omar M. Green approaches to comprehensive two-dimensional liquid chromatography (LC × LC). Journal of Chromatography Open. 2022;**2**:100046. DOI: 10.1016/j.jcoa.2022.100046

[49] Habib A, Mabrouk M, Fekry M, Mansour F. Glycerol as a novel green mobile phase modifier for reverse phase liquid chromatography. Microchemical Journal. 2021;**169**:106587. DOI: 10.1016/j.microc.2021.106587

[50] Habib A, Mabrouk M, Fekry M, Mansour F. Glycerol as a new mobile phase modifier for green liquid chromatographic determination of ascorbic acid and glutathione in pharmaceutical tablets. Journal of Pharmaceutical and Biomedical Analysis. 2022;**219**:114870. DOI: 10.1016/j.jpba.2022.114870

[51] da Silva GP, Mack M, Contiero J. Glycerol: A promising and abundant carbon source for industrial microbiology. Biotechnology Advances. 2009;**27**:30-39. DOI: 10.1016/j.biotechadv.2008.07.006

[52] Diaz-Alvarez A, Francos J, Croche P, Cadierno V. Recent advances in the use of glycerol as green solvent for synthetic organic chemistry. Current Green Chemistry. 2014;**1**:51-65. DOI: 10.2174/2213346101011131218094907

[53] Foster MD, Synovec RE. Reversed phase liquid chromatography of organic hydrocarbons with water as the mobile phase. Analytical Chemistry. 1996;**68**:2838-2844. DOI: 10.1021/ac951200+

[54] Dembek M, Bocian S. Pure water as a mobile phase in liquid chromatography

techniques. TrAC Trends in Analitical Chemistry. 2020;**123**:115793. DOI: 10.1016/j.trac.2019.115793

[55] Greibrokk T, Andersen T. High-temperature liquid chromatography. Journal of Chromatography A. 2003;**1000**:743-755. DOI: 10.1016/S0021-9673(02)01963-5

[56] Hartonen K, Riekkola M. Liquid chromatography at elevated temperatures with pure water as the mobile phase. TrAC Trends in Analytical Chemistry. 2008;**27**:1-14. DOI: 10.1016/j.trac.2007.10.010

[57] Dembek M, Bocian S. Stationary phases for green liquid chromatography. Materials. 2022;**15**:419. DOI: 10.3390/ma15020419

[58] Bocian S, Krzeminska K. The separation using pure water as a mobile phase in liquid chromatography using polar-embedded stationary phases. Green Chemistry Letters Reviews. 2019;**12**:69-78. DOI: 10.1080/17518253.2019.1576775

[59] Santos Pereira A, David F, Vanhoenacker G, Sandra P. The acetonitrile shortage: Is reversed HILIC with water an alternative for the analysis of highly polar ionizable solutes? Journal of Separation Science. 2009;**32**:2001e2007. DOI: 10.1002/jssc.200900272

[60] SIELC Technologies. 2020. Available from: https://www.sielc.com/wp-content/uploads/2020/03/CoAp-2020_2-HPLC-SEPARATION-OF-AMINO-ACIDS-IN-PURE-WATER.pdf. [Accessed: November 11, 2022]

[61] Martin A, Mundorff GH. Physical Pharmacy: Physical Chemical Principles in the Pharmaceutical Science. 4th ed. Baltimore, USA: Williams & Wilkins; 1993. pp. 396-398. ISBN: 0-8121-1428-8

[62] Unal DN, Yıldırım S, Kurbanoglu S, Uslu B. Current trends and roles of surfactants for chromatographic and electrochemical sensing. Trends in Analytical chemistry. 2021;**144**:116418. DOI: 10.1016/j.trac.2021.116418

[63] Armstrong DW, Henry SJ. Use of an aqueous micellar mobile phase for separation of phenols and polynuclear aromatic hydrocarbons via HPLC. Journal of Liquid Chromatography. 1980;**3**:657-662. DOI: 10.1080/01483918008060181

[64] Bahgat AE, Hafez MH, El-Sayed MH, Kabil NAS. Development of a solvent-free micellar HPLC method for determination of five antidiabetic drugs using response surface methodology. Microchemical Journal. 2022;**179**:107446. DOI: 10.1016/j.microc.2022.107446

[65] Rambla-Alegre M. Basic principles of MLC. Chromatography Research International. 2012;**2012**:898520. DOI: 10.1155/2012/898520

[66] Kawczak P, Bączek T. Recent theoretical and practical applications of micellar liquid chromatography (MLC) in pharmaceutical and biomedical analysis. Central European Journal of Chemistry. 2012;**10**:570-584. DOI: 10.2478/s11532-012-0004-7

[67] El-Shaheny RN, El-Maghrabey MH, Belal FF. Micellar liquid chromatography from green analysis perspectives. Open Chemistry. 2015;**13**:877-892. DOI: 10.1515/chem-2015-0101

[68] Ali AA-KF, Danielson ND. Ultra-high-performance micellar liquid chromatography comparing tween 20 and tween 40 for the determination of Hydroxycinnamic acids. Separation. 2022;**9**:61. DOI: 10.3390/separations9030061

[69] Nasr ZA, Soliman MM, Mohamed EH, Fouad FA. Assessment of the greenness of micellar HPLC method for rapid separation and simultaneous estimation of chlorpheniramine maleate in presence of some co-administrated drugs in three pharmaceutical dosage forms using single run. Acta Chromatographica. 2022;**34**:138-149. DOI: 10.1556/1326.2021.00883

[70] Ruiz-Angel M, Carda-Broch S, Torres-Lapasió JR, García-Álvarez-Coque M. Retention mechanisms in micellar liquid chromatography. Journal of Chromatography A. 2009;**1216**:1798-1814. DOI: 10.1016/j.chroma.2008.09.053

[71] Baeza-Baeza J, Dávila Y, Fernández-Navarro J, García-Alvarez-Coque M. Measurement of the elution strength and peak shape enhancement at increasing modifier concentration and temperature in RPLC. Analytical and Bioanalytical Chemistry. 2012;**404**:2973-2984. DOI: 10.1007/s00216-012-6387-7

[72] Ibrahim AE, Elmaaty A, El-Sayed H. Determination of six drugs used for treatment of common cold by micellar liquid chromatography. Analytical and Bioanalytical Chemistry. 2021;**413**:5051-5065. DOI: 10.1007/s00216-021-03469-3

[73] Patyra E, Kwiatek K. Analytical capabilities of micellar liquid chromatography and application to residue and contaminant analysis: A review. Journal of Separation Science. 2021;**44**:2206-2220. DOI: 10.1002/jssc.202001261

[74] El-Enin MAA, Salem YA, Saadia M, El-Ashry SM, Hammouda MEA. Applying eco-friendly micellar liquid chromatography for the simultaneous determination of two ternary mixtures utilized for cold treatment using monolithic column. Journal of the Chinese Chemical Society. 2021;**68**:1686-1696. DOI: 10.1002/jccs.202100093

[75] Ibrahim AE, Elmansi H, Belal F. Solvent-free mixed micellar mobile phases; an advanced green chemistry approach for reversed phase HPLC determination of some antihypertensive drugs. Journal of Separation Science. 2020;**43**:3224-3232. DOI: 10.1002/jssc.202000429

[76] Kamal AH, El-Malla SF. Mixed micellar liquid chromatographic method for simultaneous determination of norfloxacin and tinidazole in pharmaceutical tablets. Microchemical Journal. 2019;**150**:104151. DOI: 10.1016/j.microc.2019.104151

Chapter 6

Phenolic Compounds Profile of Brazilian Commercial Orange Juice

Maria Rita Estevam, Estela Mesquita and Magali Monteiro

Abstract

Orange juice is largely produced, exported and consumed in Brazil. It is an important source of bioactive compounds, such as flavonoids and phenolic acids, which are beneficial to the health of consumers. The aim of this work was to evaluate the phenolic compounds profile of commercial orange juice from Brazil using HPLC-DAD and UPLC-ESI-MS, and multivariate analysis. Forty-five phenolic compounds and one precursor were identified: cinnamic acid, 19 cinnamic acid derivatives, 8 flavanones and 18 flavones. Rutin, eriocitrin, narirutin, naringin, hesperidin, naringenin, nobiletin and tangeritin as well as caffeic, *p*-coumaric and ferulic acids were present in juices from all brands. Hesperidin and narirutin presented the highest levels considering all brands, while tangeretin and ferulic acid had the lowest levels in all juices. Principal component analysis of the phenolic compounds profile showed a wide variety among juices within the same brands, making it hard to perceive major differences among the brands.

Keywords: polyphenols, orange juice, HPLC-DAD, UPLC-ESI-MS, flavonoids, phenolic acids

1. Introduction

Brazil is the largest producer and exporter of orange juice worldwide. The crop of 2017/2018 produced 370 million boxes of oranges (40.8 kg/box) and a total of 1.3 million ton of juice. In 2019/2020, the production was of 365.4 million boxes of oranges and 965 thousand ton of juice. A total juice production of 1.2 million ton is predicted for the 2020/2021 crop. Brazilian yearly exportation was of 1.1 million ton of juice from 2018 to 2020. The majority of the juice was exported to Belgium and the Netherlands, Japan, China and the US [1].

Fresh orange juice is processed to result in Not from Concentrate (NFC) and Frozen Concentrated Orange Juice (FCOJ). While the NFC juice maintains the aroma and flavor characteristic of fresh orange juice, the FCOJ is a most desirable final product in terms of stability and storage, which is a good alternative for exportation [2, 3]. The NFC orange juice is widely available in the Brazilian market and is widely preferred by consumers when compared to FCOJ. It is also being exported alongside FCOJ. On the other hand, there is still little information concerning its composition and quality.

Orange juice is an important source of ascorbic acid and contains other important bioactive compounds such as flavonoids and phenolic acids, which have been associated to health benefits [4–6].

High Performance Liquid Chromatography (HPLC) is the main technique used for phenolic compound analysis in a wide variety of matrixes, especially foods. Reverse-phase columns, diode array detectors (HPLC-DAD) and mass spectrometer (HPLC-MS and HPLC-MS-MS) is very usual, as well as acid and polar solvents. For the separation of compounds with different structures and polarities, it is necessary to use a solvent gradient. Due to the presence of the characteristic nuclei and specific radicals, the UV/Vis spectra of flavonoids are typical for each class, usually with two bands of absorption: between 240 and 280 nm (band I, maximum absorption) and between 300 and 380 nm (band II). More specifically, flavanones present maximum absorption between 275 and 290 nm (band I) and between 300 and 380 nm (band II), with a shoulder between 300 and 340 nm (band II). Flavones, in turn, have maximum absorption between 245 and 260 nm (band I) and absorption between 350 and 380 nm (band II). The structure variations within a same class will result in discreet variations in the spectra, with hypsochromic or bathocromic shifts, although this is not sufficient for structure elucidation or differentiation. The mass spectra of flavonoids are also characteristic of the common nuclei among classes. The break of bonds produces specific fragments for glycosides and aglycones, thus making mass spectrometry a good option for structure differentiation and elucidation, and compounds identity confirmation. The same fundamentals apply for phenolic acids [7, 8].

The majority of the works report high levels of the flavonoids hesperidin and narirutin as well as lower levels of other flavanones and flavones in juice made of oranges from several varieties, including Pera-Rio [7, 9, 10]. There are fewer reports on the phenolic acid contents of orange and orange juice and to the best of our knowledge there is no available data on the phenolic composition of the Brazilian commercial orange juice. Thus, the aim of this work was to evaluate the profile of phenolic compounds of Brazilian commercial orange juice from different brands using HPLC-DAD and UPLC-ESI-MS, and employing multivariate analysis. The novelty of this work lies on the growing importance of NFC juice consumption and exportation in Brazil, which affects economic and health issues. To our knowledge, this is the first study that aims to profile the phenolic composition of commercial juice and, on top of that, evaluate this profile throughout the orange harvest. To achieve that, we applied a previously developed and validated HPLC-DAD method, alongside UPLC-ESI-MS use for identification of the compounds.

2. Material and methods

2.1 Chemicals and standards

Standards (95–99%) of gallic acid, protocatechuic acid, chlorogenic acid, caffeic acid, syringic acid, *p*-coumaric acid, ferulic acid, *trans*-cinnamic acid, rutin, eriocitrin, quercitrin, narirutin, naringin, hesperidin, quercetin, naringenin, hesperitin, nobiletin and tangeretin were obtained from Sigma Aldrich (Missouri, USA) and ChromaDex (California, USA). Standards were weighted and diluted in methanol in order to obtain standard solutions. Formic acid of analytical grade was from Synth (São Paulo, Brazil). Acetonitrile, methanol and ethyl acetate of HPLC grade were from J.T. Baker (Pennsylvania, USA), and ultrapure water was obtained from a direct Q-3 UV System from Millipore (Massachusetts, USA).

2.2 Orange juice samples

Commercial NFC orange juice from eight brands (A, B, C, D, E. F, G and H) was purchased from retail stores in the region of Araraquara, SP, Brazil. Three bottles (1 L) from each brand with distinct production dates (1, 2 and 3) between January and November of the year 2017 were acquired (n = 24). Samples were frozen and lyophilized in a Moduloyd Freeze Dryer RV8 A65413906 (Massachusetts, USA) before extraction.

2.3 Extraction of phenolic compounds

The phenolic compounds were extracted from the lyophilized juice matrix (3 g) with aqueous methanol solution (90%, v/v; 5 mL) in ultrasonic bath (20 min). Extracts were centrifuged (9000 g/20 min), supernatants were collected and the extraction was repeated once. The clean-up step using solid-phase C18 (500 mg, 6 mL) cartridges (Bond Elut, Agilent Technologies) was performed as described in a work previously published by our research group [9]. Briefly: cartridges were conditioned with ethyl acetate and aqueous formic acid (18 mL each). Juice extracts were eluted and then cartridges were washed with ethyl acetate (30 mL). Extracts were then dried under N_2 flow and suspended in aqueous formic acid (0.1% v/v), filtered through 0.22 μm cellulose disk filters and stored at −20°C until analysis. Juices were extracted in duplicate, and analyzed within a maximum of 15 days.

2.4 HPLC-DAD and UPLC-ESI-MS profiling of phenolic compounds

Profiling of phenolic compounds of Brazilian commercial orange juice was performed in two steps. At first, HPLC-DAD was used for separation, UV-Vis spectra assessment and comparison with standards (spectra and retention time). An Acquity ARC system with a diode array detector (Waters, USA) equipped with a BEH X-Bridge C18 column (250 × 4.6 mm, 5 μm) and a guard column (20 × 4.6 mm, 5 μm) was employed. Conditions were as developed and validated in the previous work of Mesquita and Monteiro [9], which were as follows: mobile phase of water:formic acid (99.9:0.1, v/v) and acetonitrile, column temperature of 50°C, flow rate of 1.0 mL.min^{-1}, and injection volume of 20 μL. The gradient was 6–10% of acetonitrile (0–16 min), 10–22% (16–36 min), 22–100% (36–38 min), and then maintained for 5 min. Column was equilibrated for 10 min between runs. The spectra were acquired from 210 to 400 nm. Orange juice extracts and standard solutions were injected in duplicate.

On the second step of the profiling an UPLC-ESI-MS was used for identification and/or confirmation of phenolic compounds identity. The system employed was the Acquity UPLC H-Class with ESI-QtoF (Xevo G2-XS QTOF Waters, USA). Conditions were adapted to the UPLC system using a BEH C18 column (2.1 × 100 mm, 1.7 μm): flow rate was of 0.4 mL min^{-1} and injection volume was of 2.0 μL. The gradient was 6–10% acetonitrile (0–10 min), 10–22% (10–24 min), 22–100% (24–24.5 min), maintained for 2.5 min. Column was equilibrated for 3 min between runs. Capillary voltage used was of 3.0 kV (ESI+) and 2.7 kV (ESI−), source temperature was 120°C and gas temperature was 400°C. The mass spectra (m/z 50–1200) were obtained in positive and negative modes. Orange juice extracts and standard solutions were injected in duplicate: once in positive and once in negative modes.

2.5 Quantification of phenolic compounds

The quantification of phenolic compounds was also performed in the HPLC-DAD system, with the conditions described in Section 2.4. Calibration curves of chlorogenic, caffeic, *p*-coumaric and ferulic acids, rutin, eriocitrin, narirutin, naringin, hesperidin, naringenin, nobiletin and tangeritin were prepared and injected in triplicate. Curve parameters are presented in **Table 1**. Orange juice extracts were injected in duplicate. Chlorogenic acid, caffeic acid, *p*-coumaric acid, ferulic acid, *trans*-cinnamic acid, rutin, eriocitrin, quercitrin, narirutin, naringin, hesperidin, naringenin, nobiletin and tangeretin were quantified using the respective standards. The phenolic compounds identified by the UV-vis spectra characteristic of chemical class and by ESI-MS were quantified and expressed as the major compound of each chemical class present in the matrix; flavanones were expressed as hesperidin, cinnamic acid derivatives as ferulic and *p*-coumaric acids; and flavones as tangeretin and rutin.

2.6 Statistical analysis

Principal component analysis (PCA) was performed based on the correlation matrix of phenolic compounds area values of all juices from each brand; 21 variables and 24 cases, respectively. Compounds were represented by name or peak number according to **Table 2**. Only the variables with a correlation ≥0.65 (rotate component matrix) were included. The PCA was performed using STATISTICA 10.0 (StatSoft, Oklahoma, USA).

Compound	Regression equation	r value	Concentration range of calibration curve (μg g^{-1})	LoD* (ng g^{-1})	LoQ* (μg)
Chlorogenic acid	y = 29101.20x + 10617.50	0.9992	0.541–21.645	—	—
Caffeic acid	y = 72900.03x − 4607.12	0.9996	0.508–2.612	22	2.51
p- coumaric acid	y = 109731.78x − 8337.45	0.9999	0.157–7.291	5	1.47
Ferulic acid	y = 67857.41x − 3178.62	0.9999	0.104–46.605	2	0.83
Rutin	y = 39706.51x − 1603.99	0.9997	0.064–9.763	1	0.71
Eriocitrin	y = 26350.73x + 1416.13	0.9999	2.845–57.053	2	1.99
Narirutin	y = 29246.56x − 1375.12	0.9999	10.024–44.330	11	2.58
Naringin	y = 36392.69x + 17919.84	0.9997	0.766–30.170	6	1.86
Hesperidin	y = 32598.58x + 9315.74	0.9999	0.221–36.290	11	7.41
Naringenin	y = 68393.50x = 30749.70	0.9994	1.158–12.587	7	0.94
Nobiletin	y = 51185.38x + 10395.22	0.9978	0.248–6.393	8	1.86
Tangeretin	y = 36293.40x + 10329.57	0.9999	0.052–80.364	17	0.91

**LoD (limit of detection) and LoQ (limit of quantification) values from Mesquita and Monteiro [9]. Chlorogenic acid was not included in the validation step.*

Table 1.
Regression equations, r values and concentration range of the calibration curves employed in the quantification of the phenolic compounds of Brazilian commercial orange juice, with limits of detection and quantification.

Peak	Compound/chemical class	Retention time (min)	Maximum wavelength absorption (nm)	$[M + H^+]$ (mass error)	$[M - H^-]$ (mass error)	Molar mass ($g\ mol^{-1}$)
1	Cinnamic acid derivative[b,c]	6.09	310; 230	—	385.1301 (3.2)	—
2	Cinnamic acid derivative[b,c]	7.45	310; 230	—	393.1502 (2.9)	—
3	Cinnamic acid derivative[b,c]	8.10	325; 215	—	399.1743 (4.3)	—
4	Cinnamic acid derivative[b,c]	8.66	310; 230	—	561.2158 (4.0)	—
5	Cinnamic acid derivative[b,c]	9.63	325; 215	—	559.1569 (3.5)	—
6	Cinnamic acid derivative[b,c]	10.06	325; 215	—	399.2172 (2.7)	—
7	Cinnamic acid derivative[b,c]	10.92	315; 220	—	355.1204 (3.3)	—
8	Chlorogenic acid[a,c]	11.33	325; 215	—	353.0867 (−1.4)	354.0946
9	Cinnamic acid derivative[b,c]	12.24	315; 230	—	331.1900 (−4.2)	—
10	Cinnamic acid derivative[b,c]	12.68	310; 230	—	325.2002 (−3.8)	—
11	Caffeic acid[a,c]	13.52	325; 215	—	179.0348 (0.3)	180.0421
12	Flavone[b,c]	13.98	330; 250	—	433.1922 (2.7)	—
13	Sinapinic acid[b,c]	15.21	325; 220	225.0763 (2.3)	223.0611 (0.7)	224.0687
14	Lucenin 2[b,c]	15.32	350; 270	611.1614 (1.1)	609.1465 (0.7)	610.1541
15	Hydroumbellic acid[b,c]	15.97	330; 215	183.0656 (2.4)	181.0506 (−1.4)	182.0584
16	Dihydroxycinnamic acid[b,c]	16.35	330; 220	—	179.0357 (−3.4)	180.0423
17	Vicenin 2[b,c]	19.94	330; 270	595.1662 (0.3)	593.1515 (0.8)	594.1593
18	Diosmetin[b,c]	20.98	355; 255	301.0718 (1.3)	299.0565 (1.8)	300.0645
19	Astragalin[b,c]	21.24	355; 255	449.1079 (0.7)	447.0932 (−0.1)	448.1006
20	Apigenol[b,c]	21.56	355; 255	—	269.0458 (0.6)	270.0531
21	*p*-coumaric acid[a,c]	21.95	310; 230	165.0543 (2.4)	163.0403 (0.3)	164.0477
22	Flavanone[b,c]	22.48	330; 285	—	714.2688 (−2.4)	—
23	Rhoifolin[b,c]	22.84	355; 255	579.1722 (1.4)	577.1568 (0.6)	578.1647
24	Flavanone[b,c]	23.84	330; 280	—	273.2879 (−3.0)	—
25	Natsudaidain[b,c]	24.58	340; 270	—	417.1197 (2.6)	418.1266

Peak	Compound/chemical class	Retention time (min)	Maximum wavelength absorption (nm)	$[M + H^+]$ (mass error)	$[M - H^-]$ (mass error)	Molar mass ($g\ mol^{-1}$)
26	Ferulic acid[a,c]	25.43	320; 215	195.0654 (2.0)	193.0506 (0.9)	194.0581
27	Isovitexin[b,c]	27.43	340; 270	433.1137 (1.3)	431.0981 (−0.2)	432.1064
28	Rutin[a,c]	27.72	355; 255	611.1613 (0.9)	609.1467 (0.8)	610.1541
29	Isoquercitrin[b,c]	28.18	330; 270	465.1033 (1.4)	463.0882 (0.8)	464.0955
30	Vicenin 1/3[b,c]	28.42	355; 255	565.1557 (0.8)	563.1407 (0.2)	564.1484
31	Eriocitrin[a,c]	28.63	330; 280	597.1823 (1.0)	595.1681 (1.7)	596.1745
32	Cinnamic acid derivative[b,c]	29.07	320; 215	—	465.2235 (2.2)	—
33	Flavanone[b,c]	32.15	330; 280	—	772.4234 (2.6)	—
34	Narirutin[a,c]	32.98	330; 280	581.1870 (0.9)	579.1728 (1.9)	580.1797
35	Naringin[a,c]	33.28	330; 280	—	579.1724 (0.9)	580.1797
36	Cinnamic acid derivative[b,c]	34.93	330; 215	—	297.1789 (1.9)	—
37	Hesperidin[a,c]	36.25	335; 280	611.1982 (0.9)	609.1828 (0.3)	610.1902
38	Cinnamic acid[b,c]	36.48	265	149.0597 (−0.2)	147.0446 (−3.6)	148.0524
39	Cinnamic acid derivative[b,c]	37.28	330; 215	—	297.1766 (−2.0)	—
40	Naringenin[a,c]	39.99	330; 280	273.0758 (0.2)	271.0613 (0.3)	272.0686
41	Diosmin[b,c]	40.79	325; 240	609.1822 (1.4)	607.1675 (1.7)	608.1749
42	3.5.6-trihydroxy-3′.4′.7-trimethoxyflavone[b,c]	40.91	325; 250	361.0924 (2.3)	359.0781 (1.6)	360.0851
43	Nobiletin[a,c]	41.01	340; 250	403.1392 (0.3)	—	402.1319
44	Heptamethoxyflavone[b,c]	41.13	340; 250	303.0869 (2.0)	301.0724 (3.5)	302.0796
45	Tangeritin[a,c]	41.21	315; 270	373.1287 (0.4)	—	372.1214
46	3.5.8-trihydroxy-3′.4′-dimethoxyflavone[b,c]	41.50	325; 250	331.0809 (−1.0)	329.0664 (−2.4)	330.0737

[a]*compounds identified by retention time and UV-vis spectra compared to pure analytical standards;*[b]*compounds identified by chemical class based on UV-vis spectra;*[c]*compounds identified by UPLC-ESI-MS.*

Table 2.
Compounds identified in Brazilian commercial orange juice.

3. Results and discussion

3.1 Phenolic compounds profile

The typical phenolic compounds profile of Brazilian commercial orange juice obtained with HPLC-DAD is presented in **Figure 1**. The chromatographic profile was similar among juices from all brands, considering all production dates.

Table 2 presents the compounds identified in the juices; peaks were numbered according to retention time (1–46). The identification was based on: (1) retention time from HPLC-DAD separation and UV-vis spectra of compounds compared to pure analytical standards, (2) UV-vis spectra characteristic of flavonoids or phenolic acid chemical classes, and (3) molar mass obtained by UPLC-ESI-MS. Forty six compounds were identified: 26 flavonoids and 19 phenolic acids, as well as the phenolic acid precursor cinnamic acid (**Table 2**). The majority of the compounds were identified in all brands evaluated, considering all production dates.

All phenolic acids identified in Brazilian commercial orange juice were cinnamic acid derivatives, as verified by the UV-vis spectra characteristic of this chemical class. The spectra showed absorption from 270 to 370 nm, with a band of high intensity from 300 to 370 nm, a band of lower intensity from 270 to 300 nm and maximum absorption between 325 and 330 nm [9, 11]. Sinapinic acid (peak 13), hydroumbellic acid (peak 15), dihydroxycinnamic acid (peak 16), and *p*-coumaric acid (peak 21), as well as 11 unnamed derivatives (peaks 2–7, 9, 10, 32, 36 and 39) were present in juices from all brands, in varying levels. Chlorogenic acid (peak 8), caffeic acid (peak 11), ferulic acid (peak 26), and one unnamed derivative (peak 1) were also identified, but not in all brands.

Quantified levels of all compounds are shown in **Tables 3** and **4**. **Table 3** shows the range of concentration and **Table 4** shows the concentration of all the compounds in the brands evaluated, considering all production dates. The highest level was of the derivative represented by peak 7 (7.98 mg 100 g^{-1}), whereas the lowest was of ferulic

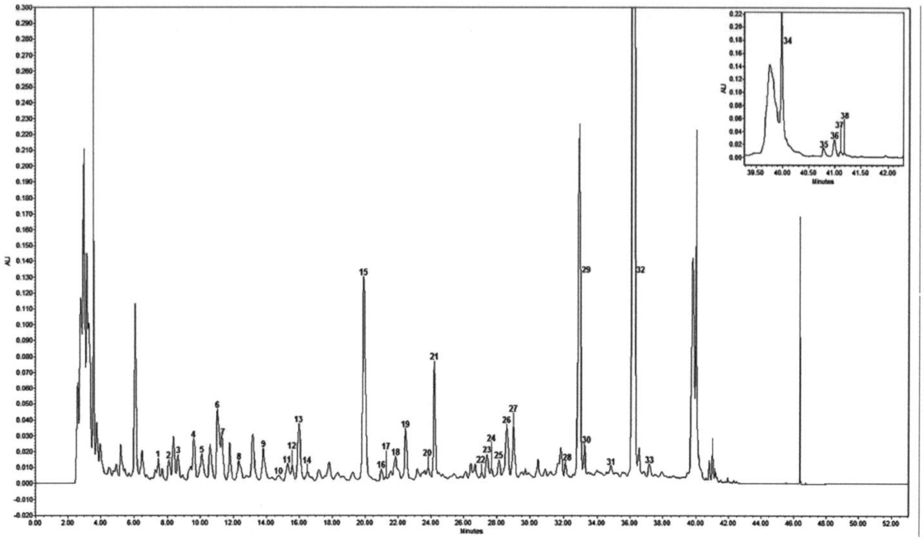

Figure 1.
Typical HPLC-DAD chromatogram of Brazilian commercial orange juice. Conditions as described in Section 2.4. Peaks numbered according to ***Table 1****.*

Peak	Compound/chemical class	Range of concentration (mg 100 g^{-1} of lyophilized juice)							
		A	B	C	D	E	F	G	H
1	Cinnamic acid derivative	0.16–0.24	nd	nd	0.17–0.27	0.10–0.23	0.13–0.21	0.16–0.51	nd
2	Cinnamic acid derivative	0.24–0.33	0.27–0.75	0.13–0.46	0.20–0.39	0.11–0.25	0.14–0.39	0.32–0.45	0.14–0.61
3	Cinnamic acid derivative	0.18–0.29	0.26–0.52	0.11–0.51	0.27–0.35	0.15–0.44	0.13–0.31	0.27–0.51	0.25–0.46
4	Cinnamic acid derivative	0.36–0.47	0.39–1.17	0.17–0.54	0.32–0.59	0.21–0.51	0.37–0.60	0.42–0.64	0.21–0.81
5	Cinnamic acid derivative	1.11–1.88	0.76–1.17	0.43–2.71	0.65–0.95	0.15–0.86	1.61–1.79	0.92–1.19	0.52–0.89
6	Cinnamic acid derivative	nd–0.66	0.13–0.73	nd–0.06	nd–0.30	0.07–3.21	nd–0.59	nd–1.03	0.58–0.99
7	Cinnamic acid derivative	3.28–4.34	1.15–1.56	1.76–4.37	1.47–2.52	2.71–4.25	2.61–7.98	1.97–3.78	1.06–2.11
8	Chlorogenic acid	nd–0.35	nd–1.31	nd	0.53–0.75	nd	nd	nd–1.95	0.56–1.29
9	Cinnamic acid derivative	0.56–0.65	0.31–0.70	0.22–0.53	0.33–0.74	0.25–0.52	0.26–0.74	0.45–0.57	0.36–0.69
10	Cinnamic acid derivative	0.08–0.21	0.09–0.12	0.06–0.18	0.07–0.19	0.06–0.08	0.13–0.22	nd–0.17	nd–0.08
11	Caffeic acid	nd–0.43	nd	nd–1.40	0.18–0.41	nd	nd - 0.34	nd–0.79	nd–0.44
12	Flavone	nd–1.48	0.90–2.24	0.44–2.08	0.15–1.82	1.00–3.27	0.08–2.91	0.27–2.89	0.94–1.64
13	Sinapinic acid	0.22–0.72	0.34–0.61	0.20–0.86	nd–0.64	0.22–0.47	0.27–0.59	nd–0.57	0.31–0.54
14	Lucenin 2	0.73–0.95	0.36–1.33	0.31–0.90	0.07–0.57	0.54–1.73	0.37–1.99	nd–1.13	0.66–1.63
15	Hydroumbellic acid	1.77–1.92	1.15–1.83	0.78–3.13	1.47–2.45	0.92–2.05	0.41–2.95	1.46–1.83	1.39–4.77
16	Dihydroxycinnamic acid	0.11–0.28	0.17–0.25	0.11–0.35	0.06–0.28	0.13–0.30	0.15–0.43	0.13–0.38	0.14–0.33
17	Vicenin 2	9.87–17.02	8.33–15.72	3.61–19.18	6.24–16.24	5.26–21.00	4.86–24.20	13.09–19.10	7.27–10.80
18	Diosmetin	1.52–1.65	0.87–1.20	0.47–2.02	0.94–1.59	0.64–2.14	1.24–1.90	1.89–1.94	0.73–1.50
19	Astragalin	0.31–1.20	0.24–0.35	0.11–0.47	0.13–0.29	0.16–0.42	0.19–0.53	0.31–0.34	0.15–0.37
20	Apigenol	0.26–0.33	0.17–0.34	0.07–0.46	0.02–0.14	0.19–0.77	0.10–0.46	0.40–0.88	0.16–0.46
21	*p*-coumaric acid	0.30–0.80	0.13–0.30	0.12–0.21	0.19–0.36	0.10–0.19	0.47–0.84	0.05–0.19	0.08–0.19
22	Flavanone	2.19–3.56	2.01–3.72	0.92–4.40	0.98–1.92	1.39–4.42	2.45–4.69	3.34–4.24	2.02–3.50
23	Rhoifolin	0.38–0.54	0.25–0.60	0.11–0.60	0.44–2.59	0.15–0.44	0.37–0.50	0.58–0.88	0.16–0.37
24	Flavanone	2.62–5.05	3.56–28.35	1.46–7.20	2.74–3.86	2.35–10.25	3.46–6.74	6.64–9.69	2.79–5.56
25	Natsudaidain	0.19–0.28	0.12–0.32	0.06–0.37	0.15–0.20	0.15–0.46	0.25–0.40	0.25–0.33	0.17–0.31

Peak	Compound/chemical class	Range of concentration (mg 100 g^{-1} of lyophilized juice)							
		A	B	C	D	E	F	G	H
26	Ferulic acid	nd–0.74	nd- 0.12	nd–0.07	0.05–0.10	nd–0.48	0.09–0.10	nd	nd–0.04
27	Isovitexin	0.87–1.58	0.66–1.75	0.28–1.92	0.78–1.21	0.57–2.35	0.96–1.93	1.92–2.33	0.88–1.74
28	Rutin	0.27–0.99	0.14–0.68	0.15–0.58	0.25–0.28	0.35–1.00	0.37–1.47	0.61–0.76	0.45–0.89
29	Isoquercitrin	0.68–1.14	0.66–1.54	0.26–1.65	0.66–0.77	0.51–1.94	0.65–1.60	1.52–1.85	0.68–1.40
30	Vicenin 1/3	0.45–0.52	0.25–0.42	0.11–0.84	0.25–0.46	0.20–0.54	0.34–0.52	0.42–0.48	0.14–0.30
31	Eriocitrin	3.72–6.11	3.14–5.18	1.51–7.36	2.62–3.90	2.89–10.47	3.72–8.25	5.43–7.39	1.90–5.21
32	Cinnamic acid derivative	0.80–1.34	1.07–2.18	0.37–3.62	0.84–1.29	0.76–2.65	0.97–1.83	2.07–2.41	0.94–1.91
33	Flavanone	0.38–1.35	0.37–0.53	0.19–1.12	0.22–0.56	0.08–0.77	0.61–1.70	0.77–1.21	0.35–0.93
34	Narirutin	8.22–17.60	8.15–19.80	4.45–24.91	8.76–15.82	7.62–36.76	5.21–30.11	19.86–28.56	8.02–13.76
35	Naringin	1.75–1.84	0.92–1.54	0.50–2.17	0.51–1.18	0.88–2.09	1.61–3.01	1.53–2.10	0.69–5.39
36	Cinnamic acid derivative	0.19–0.38	0.17–0.27	0.09–0.56	0.14–0.21	0.14–0.43	0.18–0.43	0.28–0.44	0.15–0.25
37	Hesperidin	60.94–99.89	41.80–62.51	29.20–113.17	46.27–122.00	23.21–88.85	49.79–97.45	42.25–72.44	30.14–44.21
38	Cinnamic acid	nd	nd	nd–0.94	nd	nd–0.54	nd	nd	nd
39	Cinnamic acid derivative	0.24–0.38	0.25–0.51	0.10–0.79	0.20–0.34	0.17–0.53	0.19–0.44	0.53–0.57	0.26–0.47
40	Naringenin	0.54–1.09	0.65–2.15	0.34–2.00	1.32–1.73	0.44–2.22	0.64–1.19	1.98–2.73	0.80–1.84
41	Diosmin	0.15–0.37	0.21–0.51	0.05–0.39	0.24–0.63	0.13–0.51	0.09–0.32	0.34–0.50	0.07–0.31
42	3.5.6-trihydroxy-3′.4′.7-trimethoxyflavone	<LoQ	<LoQ	<LoQ	<LoQ	<LoQ	<LoQ	<LoQ	<LoQ
43	Nobiletin	0.26–0.38	0.26–0.72	0.10–0.54	0.42–0.75	0.16–0.65	0.23–0.36	0.52–0.68	0.28–0.65
44	Heptamethoxyflavone	0.12–0.18	0.13–0.38	0.06–0.22	0.13–0.39	0.05–0.21	0.13–0.17	0.16–0.24	0.11–0.25
45	Tangeritin	<LoQ	<LoQ–0.02	<LoQ	0.01–0.03	<LoQ–0.02	<LoQ–0.02	0.01	<LoQ–0.01
46	3.5.8-trihydroxy-3′.4′-dimethoxyflavone	0.02–0.04	nd–0.02	nd	nd–0.07	nd	nd–0.05	nd	nd

Results expressed as mean value.
nd: not detected; LoQ: limit of quantification [9].

Table 3.
Phenolic compounds (mg 100 g^{-1}) of Brazilian commercial orange juice.

Peak	Compound/chemical class	Concentration (mg 100 g^{-1})																							
		A			B			C			D			E			F			G			H		
		1	2	3	1	2	3	1	2	3	1	2	3	1	2	3	1	2	3	1	2	3	1	2	3
1	Cinnamic acid derivative	0.24	0.19	0.16	nd	nd	nd	nd	nd	nd	0.17	0.19	0.27	0.14	0.23	0.10	0.13	0.18	0.21	0.16	0.28	0.51	0.16	0.28	0.51
2	Cinnamic acid derivative	0.24	0.27	0.33	0.75	0.31	0.27	0.13	0.46	0.34	0.20	0.22	0.39	0.25	0.21	0.11	0.14	0.27	0.39	0.32	0.39	0.45	0.32	0.39	0.45
3	Cinnamic acid derivative	0.29	0.22	0.18	0.52	0.27	0.26	0.11	0.33	0.51	0.30	0.27	0.35	0.44	0.33	0.15	0.13	0.22	0.31	0.27	0.51	0.51	0.27	0.51	0.51
4	Cinnamic acid derivative	0.47	0.36	0.38	1.17	0.49	0.39	0.17	0.54	0.35	0.32	0.36	0.59	0.51	0.36	0.21	0.37	0.48	0.60	0.42	0.56	0.64	0.42	0.56	0.64
5	Cinnamic acid derivative	1.88	1.11	1.19	1.17	1.05	0.76	0.43	1.85	2.71	0.95	0.65	0.86	0.15	0.86	0.56	1.61	1.79	1.66	1.10	1.19	0.92	1.10	1.19	0.92
6	Cinnamic acid derivative	0.66	nd	0.58	0.73	0.16	0.13	0.06	nd	nd	0.14	nd	0.30	0.17	3.21	0.07	0.54	0.59	nd	nd	0.54	1.03	nd	0.54	1.03
7	Cinnamic acid derivative	4.34	3.94	3.28	1.56	1.56	1.15	1.76	4.37	3.03	2.52	1.72	1.47	4.25	2.71	2.71	2.61	4.93	7.98	3.78	2.90	1.97	3.78	2.90	1.97
8	Chlorogenic acid	0.35	nd	nd	1.31	nd	nd	nd	nd	nd	0.61	0.53	0.75	nd	nd	nd	nd	nd	nd	nd	1.29	1.95	0.63	0.56	1.29
9	Cinnamic acid derivative	0.56	0.60	0.65	0.70	0.55	0.31	0.23	0.53	0.22	0.34	0.33	0.74	0.52	0.38	0.25	0.26	0.59	0.74	0.57	0.57	0.45	0.57	0.57	0.45
10	Cinnamic acid derivative	0.08	0.17	0.21	0.12	0.11	0.09	0.06	0.18	0.08	0.12	0.07	0.19	0.06	0.08	0.07	0.13	0.21	0.22	0.17	0.11	nd	0.17	0.11	nd
11	Caffeic acid	0.43	nd	nd	nd	nd	nd	nd	nd	1.40	0.18	0.41	nd	nd	nd	nd	0.34	0.29	nd	nd	0.79	nd	0.44	nd	nd
12	Flavone	nd	1.48	1.18	2.24	1.27	0.90	0.44	1.83	2.08	1.69	0.15	1.82	3.27	1.55	1.00	0.08	0.35	2.91	2.89	0.27	2.36	2.89	0.27	2.36
13	Sinapinic acid	0.22	0.72	0.51	0.61	0.41	0.34	0.20	0.66	0.86	0.46	nd	0.64	0.47	0.28	0.22	0.27	0.46	0.59	0.57	nd	0.53	0.57	nd	0.53
14	Lucenin 2	0.73	0.80	0.95	1.33	0.64	0.36	0.31	0.77	0.90	0.07	0.11	0.57	1.73	1.02	0.54	0.37	1.19	1.99	1.13	nd	0.94	1.13	nd	0.94
15	Hydroumbellic acid	1.92	1.83	1.77	1.71	1.83	1.15	0.78	2.03	3.13	1.55	1.47	2.45	2.05	1.52	0.92	0.41	1.89	2.95	1.83	1.75	1.46	1.83	1.75	1.46
16	Dihydroxycinnamic acid	0.11	0.28	0.27	0.25	0.20	0.17	0.11	0.35	0.32	0.10	0.06	0.28	0.30	0.15	0.13	0.15	0.20	0.43	0.38	0.13	0.27	0.38	0.13	0.27
17	Vicenin 2	17.02	9.87	11.93	15.72	8.33	10.68	3.61	16.08	19.18	6.24	6.34	16.24	21.00	14.92	5.26	4.86	11.85	24.20	19.10	18.75	13.09	19.10	18.75	13.09

Peak	Compound/chemical class	Concentration (mg 100 g^{-1})																							
		A			B			C			D			E			F			G			H		
		1	2	3	1	2	3	1	2	3	1	2	3	1	2	3	1	2	3	1	2	3	1	2	3
18	Diosmetin	1.52	1.54	1.65	1.20	0.97	0.87	0.47	1.35	2.02	1.00	0.94	1.59	2.14	1.16	0.64	1.24	1.64	1.90	1.94	1.89	1.91	1.94	1.89	1.91
19	Astragalin	1.20	0.34	0.31±	0.35	0.24	0.26	0.11	0.38	0.47	0.13	0.13	0.29	0.42	0.27	0.16	0.19	0.37	0.53	0.32	0.31	0.34	0.32	0.31	0.34
20	Apigenol	0.32	0.33	0.26	0.17	0.21	0.34	0.07	0.46	0.46	0.02	0.06	0.14	0.77	0.56	0.19	0.10	0.34	0.46	0.40	0.49	0.88	0.40	0.49	0.88
21	*p*-coumaric acid	0.80	0.31	0.30	0.13	0.30	0.13	0.12	0.21	0.16	0.25	0.19	0.36	0.19	0.11	0.10	0.84	0.57	0.47	0.19	0.10	0.05	0.19	0.08	0.08
22	Flavanone	3.56	2.96	2.19	3.72	2.35	2.01	0.92	3.47	4.40	1.92	1.48	0.98	4.42	2.46	1.39	2.45	2.96	4.69	4.24	3.92	3.34	4.24	3.92	3.34
23	Rhoifolin	0.54	0.44	0.38	0.60	0.31	0.25	0.11	0.42	0.60	0.44	0.48	2.59	0.44	0.26	0.15	0.37	0.48	0.50	0.58	0.73	0.88	0.58	0.73	0.88
24	Flavanone	5.05	3.30	2.62	5.03	28.35	3.56	1.46	5.51	7.20	3.86	2.84	2.74	10.25	4.99	2.35	3.46	4.04	6.74	9.69	8.39	6.64	9.69	8.39	6.64
25	Natsudaidain	0.28	0.25	0.19	0.32	0.12	0.13	0.06	0.20	0.37	0.20	0.15	0.15	0.46	0.24	0.15	0.25	0.31	0.40	0.33	0.30	0.25	0.33	0.30	0.25
26	Ferulic acid	0.74	0.51	nd	nd	0.12	nd	0.04	0.07	nd	0.06	0.05	0.10	0.48	nd	nd	0.09	0.10	0.09	nd	nd	nd	0.04	nd	nd
27	Isovitexin	1.58	0.93	0.87	1.75	1.01	0.66	0.28	1.22	1.92	0.78	1.02	1.21	2.35	1.13	0.57	0.96	1.21	1.93	1.92	2.14	2.33	1.92	2.14	2.33
28	Rutin	0.99	0.37	0.27	0.68	0.40	0.14	0.15	0.23	0.58	0.28	0.27	0.25	1.00	0.69	0.35	0.37	1.14	1.47	0.61	0.76	0.69	0.45	0.47	0.89
29	Isoquercitrin	1.14	0.74	0.68	1.54	0.75	0.66	0.26	1.07	1.65	0.66	0.70	0.77	1.94	1.16	0.51	0.65	1.02	1.60	1.52	1.73	1.85	1.52	1.73	1.85
30	Vicenin 1/3	0.45	0.52	0.50	0.42	0.28	0.25	0.11	0.44	0.84	0.25	0.25	0.46	0.54	0.27	0.20	0.34	0.46	0.52	0.42	0.46	0.48	0.42	0.46	0.48
31	Eriocitrin	6.11	4.04	3.72	5.18	4.27	3.14	1.51	5.23	7.36	2.62	2.69	3.90	10.47	4.65	2.89	3.72	5.19	8.25	7.39	6.33	5.43	2.86	1.90	5.21
32	Cinnamic acid derivative	1.34	1.02	0.80	2.18	1.07	1.18	0.37	1.81	3.62	1.29	0.84	0.84	2.65	1.40	0.76	0.97	1.19	1.83	2.07	2.41	2.18	2.07	2.41	2.18
33	Flavanone	1.35	0.49	0.38	0.41	0.37	0.53	0.19	1.12	1.11	0.56	0.37	0.22	0.08	0.77	0.43	0.61	0.71	1.70	1.21	0.86	0.77	1.21	0.86	0.77
34	Narirutin	17.60	10.33	8.22	19.80	8.15	12.99	4.45	13.97	24.91	8.76	13.48	15.82	36.75	22.61	7.62	5.21	9.71	30.11	28.56	27.02	19.86	8.68	13.76	8.02
35	Naringin	1.78	1.84	1.75	1.54	0.92	1.32	0.50	2.17	1.97	0.51	0.64	1.18	2.09	1.58	0.88	1.61	2.48	3.01	2.10	1.76	1.53	0.69	0.69	5.39
36	Cinnamic acid derivative	0.38	0.25	0.19	0.27	0.19	0.17	0.09	0.27	0.56	0.21	0.14	0.16	0.43	0.29	0.14	0.18	0.29	0.43	0.44	0.36	0.28	0.44	0.36	0.28
37	Hesperidin	99.89	78.04	60.94	62.51	41.80	47.12	29.20	65.45	113.17	46.27	53.98	122.0C	88.85	46.11	23.21	49.79	71.80	97.45	72.44	64.45	42.25	30.14	44.21	38.84
38	Cinnamic acid	nd	nd	nd	nd	nd	nd	nd	nd	0.94	nd	nd	nd	0.54	nd	nd	nd	nd	nd	nd	nd	nd	nd	nd	nd
39	Cinnamic acid derivative	0.38	0.28	0.24	0.51	0.27	0.25	0.10	0.40	0.79	0.25	0.20	0.34	0.53	0.32	0.17	0.19	0.32	0.44	0.53	0.57	0.54	0.53	0.57	0.54

Peak	Compound/chemical class	Concentration (mg 100 g^{-1})																							
		A			B			C			D			E			F			G			H		
		1	2	3	1	2	3	1	2	3	1	2	3	1	2	3	1	2	3	1	2	3	1	2	3
40	Naringenin	1.09	0.74	0.54	2.15	0.65	0.80	0.34	1.67	2.00	1.32	1.47	1.73	2.22	0.99	0.44	0.64	0.36	1.19	1.98	2.01	2.73	1.39	0.80	1.84
41	Diosmin	0.37	0.15	0.16	0.51	0.26	0.21	0.05	0.32	0.39	0.47	0.24	0.63	0.51	0.25	0.13	0.09	0.14	0.32	0.36	0.50	0.34	0.36	0.50	0.34
42	3.5.6-trihydroxy-3′.4′.7-trimethoxyflavone	< LoQ	< LoQ	< LoQ	< LoQ	< LoQ	< LoQ	< LoQ	< LoQ	< LoQ	< LoQ	< LoQ	<LoQ	< LoQ	< LoQ	< LoQ	< LoQ	< LoQ	< LoQ	< LoQ	< LoQ	< LoQ	< LoQ	< LoQ	< LoQ
43	Nobiletin	0.38	0.30	0.26	0.72	0.40	0.26	0.10	0.39	0.54	0.67	0.42	0.75	0.65	0.36	0.16	0.23	0.36	0.36	0.52	0.67	0.68	0.42	0.28	0.65
44	Heptamethoxyflavone	0.18	0.16	0.12	0.38	0.21	0.13	0.06	0.22	0.18	0.13	0.14	0.39	0.21	0.12	0.05	0.13	0.17	0.16	0.16	0.20	0.24	0.16	0.20	0.24
45	Tangeritin	< LoQ	< LoQ	< LoQ	0.02	0.02	< LoQ	< LoQ	< LoQ	< LoQ	0.03	0.01	0.01	0.02	0.01	< LoQ	< LoQ	0.02	0.02	0.01	0.01	0.01	0.01	< LoQ	< LoQ
46	3.5.8-trihydroxy-3′.4′-dimethoxyflavone	0.04	0.03	0.02	nd	0.02	nd	nd	nd	nd	nd	nd	0.07	nd	nd	nd	0.05	0.04	nd	nd	nd	nd	0.16	0.28	0.51

Results expressed as mean value; standard deviations varied from ±0.00 to ±2.19.
LoQ (limit of quantification) according to ***Table 3****.*
nd: not detected.

Table 4.
Concentration of phenolic compounds (mg 100 g^{-1}) of Brazilian commercial orange juice.

acid (0.04 mg 100 g^{-1}). The precursor cinnamic acid was also found in low levels (0.54–0.94 mg 100 g^{-1}) in two of the brands evaluated, C and E (**Table 3**). Chlorogenic acid (0.3–2 mg 100 g^{-1}), caffeic acid (0.2–1 mg 100 g^{-1}) and ferulic acid (0.04–0.7 mg 100 g^{-1}) were present only in some of the brands evaluated, whereas *p*-coumaric acid (0.1–0.8 mg 100 g^{-1}) was in all of them.

These results were compared with those reported in the work of Mesquita and Monteiro [9], which was developed previously. The authors evaluated organic and conventional orange juice throughout the 2016 harvest. All of the phenolic acids identified by them were cinnamic acid derivatives like in this work, although we could not confirm if they were the same ones. Also, a greater number of phenolic acids (19 compounds, and the precursor cinnamic acid) were identified in the present work, whereas only 7 was previously reported [9].

Among the flavonoids identified in Brazilian commercial orange juice were 8 flavanones: eriocitrin (peak 31), narirutin (peak 34), naringin (peak 35), hesperidin (peak 37), naringenin (40) and 3 compounds with no matches in MS library (peaks 22, 24 and 33). These compounds showed UV-vis spectra characteristic of this class: absorption from 250 to 360 nm with a band of high intensity (250–300 nm), maximum absorption between 280 and 285 nm and a shoulder from 300 to 360 nm [8, 9]. Regarding the flavones, 18 compounds were identified: lucenin-2 (peak 14), vicenin-2 (peak 17), diosmetin (peak 18), astragalin (peak 19), apigenol (20), rhoifolin (peak 23), natsudaidain (peak 25), isovitexin (peak 27), rutin (peak 28), isoquercitrin (peak 29), vicenin 1 or 3 (peak 30), diosmin (peak 41), 3.5.6-trihydroxy-3′.4′.7-trimethoxyflavone (peak 42), nobiletin (peak 43), heptamethoxyflavone (peak 44), tangeretin (peak 45) and 3.5.8-trihydroxy-3′.4′-dimethoxyflavone (peak 46). Only the compound represented by peak 12 had no match in the MS library. It showed UV-vis spectra corresponding to the flavones class (200–400 nm, band from 230 to 280 nm and maximum absorption at 255 nm) [9, 12].

The majority of the flavonoids were identified in all of the brands evaluated, with the exception only of naringenin (not identified in brand B), lucenin 2 (identified only in brands D, E and H), heptamethoxyflavone (identified in brands A, C, D, E and G) and 3,5,8-trihydroxy-3′,4′-dimethoxyflavone (identified in A, B, C, F, G and H).

The most expressive compounds of the juice were hesperidin (23–122 mg 100 g^{-1}) and narirutin (4–37 mg 100 g^{-1}), respectively. Intermediate levels of eriocitrin (1–10 mg 100 g^{-1}) and naringin (0.5–5 mg 100 g^{-1}) were also observed. Naringenin (0.3–3 mg 100 g^{-1}), rutin (0.1–1 mg 100 g^{-1}) and nobiletin (0.1–0.8 mg 100 g^{-1}) levels were lower. Tangeretin was present in even lower levels (0.01–0.03 mg 100 g^{-1}), sometimes below the limit of quantification (0.91 μg) [9] (**Table 3**).

Orange juice is widely consumed worldwide; however, there are few works that report the profile of phenolic compounds of this matrix. The most reported flavonoids in orange and orange juice are the flavanones hesperidin and narirutin [7, 10]. Mesquita and Monteiro [9] identified rutin, hesperidin, narirutin, eriocitrin, nobiletin and tangeritin in organic and conventional Pêra-Rio orange juices from Brazil throughout the 2016 harvest. All of the compounds identified and confirmed by them were also identified in juices from all brands evaluated in this work. Besides those compounds, the authors also found 12 other flavonoids they could not identify (6 flavones and 6 flavanones). In this work, the majority of the flavonoids were identified by MS, even when standards were not available. Brazilian commercial orange juice showed levels lower than NFC, FCOJ and freshly-squeezed organic and conventional orange juices as reported by Mesquita and Monteiro [9], considering the compounds evaluated.

3.2 Principal component analysis

For principal component analysis (PCA), phenolic compounds were represented by name and/or peak number according to **Table 2**. Only the variables with a correlation ≥0.65 were used; chlorogenic acid, nobiletin, naringenin, isovitexin, isoquercitrin, apigenol, narirutin, diosmetin, natsudaidain, vicenin 2, vicenin 1/3, rutin, hesperidin, astragalin, eriocitrin, and compounds represented by peaks 3, 7, 22, 32, 36 and 39.

The PCA model was calculated on the autoscaled data (21 variables, 24 samples). The first two principal components (PC 1 and PC 2) explained a variation of 80.36% in the levels of phenolic compounds of the orange juices. PCA enabled the discrimination of the juice from different brands and production dates according to spatial distribution. As seen on **Figure 2a**, all compounds were loaded negatively in PC 1. Chlorogenic acid, nobiletin, naringenin, isovitexin, isoquercitrin and apigenol, and compounds correspondent to peaks 3, 32 and 39 were loaded positively in PC 2. Narirutin was loaded on zero in PC 2. Diosmetin, vicenin 2, vicenin 1/3, natsudaidain, rutin, hesperidin, eriocitrin, astragalin and compounds correspondent to peaks 7, 22 and 36 were loaded negatively in PC2 (**Figure 2a**).

Figure 2b shows the plot distribution of juice samples from different brands (A, B, C, D, E, F, G and H) and production dates (1, 2 and 3). Overall, there was a difference in the levels of phenolic compounds of juices within a same brand for most of the brands evaluated. All juices from G were loaded negatively in PC 1; G1 was loaded negatively in PC2 as well, whereas G2 and G3 were loaded positively. Juices from B, D and H were all loaded positively in PC 2: D1, D2 and D3 were loaded together, showing that this brand was the one with the least difference among production dates. Juices from brand B (B2 and B3) were loaded positively in PC 1 and also close together, whereas B1 was loaded negatively in PC1. Juices from brand A (A1, A2, A3) showed a similar profile and were loaded closer. Juices from brands C, E and F showed the greatest differences among production dates. Overall, G1, G2, G3, C3, E1 and F3 were the juices with the highest levels of expressive phenolic compounds.

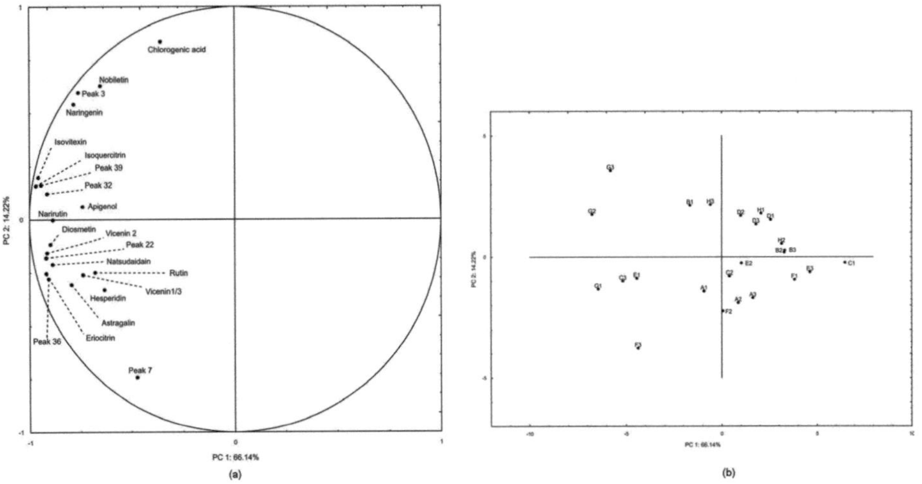

Figure 2.
*Principal component analysis of phenolic compounds (named according to **Table 1**) (a) from Brazilian commercial orange juices from 8 brands (A, B, C, D, E, F, G and H), and from different production dates (1, 2 and 3) (b).*

The presented results showcase that PCA allowed the differentiation of the Brazilian commercial orange juice according to differences and similarities in the phenolic compounds profile. A wide variety was observed among the juices within the same brands, as opposed to among the brands. This great variation was observed in the majority of the brands, making it difficult to verify any major differences among brands.

Considering that juices were acquired from January to November 2017, we can assume that the oranges used were in different maturity degrees, since the orange crop harvest is from July to October. It is well known that the levels of secondary metabolites (such as the phenolic compounds) vary as the fruit matures throughout the harvest. Thus, it is understandable that juices show this difference in the profile.

4. Conclusions

The chromatographic profile of phenolic compounds in Brazilian commercial orange juice from 8 brands was evaluated. Forty-six compounds were identified and quantified: 26 flavonoids, 19 phenolic acids, and the precursor cinnamic acid. All phenolic acids identified were cinnamic acid derivatives, as confirmed by MS and/or UV-vis spectra. Regarding the flavonoids, 8 of the compounds were flavanones, and 18 were flavones. The most expressive levels in all juices were of hesperidin and narirutin, respectively. The lowest levels were of ferulic acid and tangeretin, which sometimes was even lower than the limit of quantification. PCA was able to showcase the differences and similarities of commercial orange juice from different brands according to the influence of phenolic compounds, considering different production dates.

This work showcases the importance of the phenolic compounds profile and how significantly it can influence the juice. We observed that even within a same brand, there were great differences among juices produced. This can be related to the maturity of the fruits and harvest period. Finally, this work assessed the quality of the NFC juice, which is widely consumed in Brazil and exported worldwide. To our knowledge, there are no other works that evaluate the profile of these juices. It is an important work to base new research aiming to further evaluate the contents of orange juice.

Acknowledgements

This work was funded by São Paulo Research Foundation (FAPESP) (Processes n° 2013/10138-0 and 2014/23303-1) and National Council for Scientific and Technological Development—CNPq (scholarship grant of author M.R. Estevam). Authors would also like to thank Waters Technologies (São Paulo, SP, Brazil) for the support with UPLC-ESI-MS analysis.

Author details

Maria Rita Estevam, Estela Mesquita and Magali Monteiro*
Department of Food and Nutrition, School of Pharmaceutical Science, São Paulo State University–UNESP, Araraquara, SP, Brazil

*Address all correspondence to: magali.monteiro@unesp.br

References

[1] Brazil: Citrus Annual 2020 [Online]. U.S. Department of Agriculture; 2020. Available from: https://www.fas.usda.gov/data/brazil-citrus-annual-3 [March 3, 2021]

[2] Mastello RB, Janzantti NS, Monteiro M. Volatile and odoriferous compounds changes during frozen concentrated orange juice processing. Food Research International. 2015;**77**:591-598

[3] Janzantti NS, Machado TV, Monteiro M. Sensory acceptance of juice from FCOJ processing steps. Journal of Sensory Studies. 2011;**26**: 322-330

[4] Di Majo D, Giammanco M, La Guardia M, Tripoli E, Giammanco S, Finotti E. Flavanones in Citrus fruit: Structure-antioxidant activity relationships. Food Research International. 2015;**38**:161-1166

[5] Havsteen BH. The biochemistry and medical significance of the flavonoids. Pharmacology & Therapeutics. 2012;**96**: 67-202

[6] Roza JM, Xian-Liu Z, Guthrie N. Effect of citrus flavonoids and tocotrienols on serum cholesterol levels in hypercholesterolemic subjects. Alternative Therapies in Health and Medicine. 2007;**13**:44-48

[7] Gattuso G, Barreca D, Gargiulli C, Leuzzi U, Caristi C. Flavonoid composition of citrus juices. Molecules. 2007;**12**:1641-1673

[8] Barreca D, Gattuso G, Laganà G, Leuzzi U, Belloco E. C- and O-glycosyl flavonoids in Sanguinello and Tarocco blood orange (Citrus sinensis (L.) Osbeck) juice: Identification and influence on antioxidant properties and acetylcholinesterase activity. Food Chemistry. 2016;**196**:619-627

[9] Mesquita E, Monteiro M. Simultaneous HPLC determination of flavonoids and phenolic acids profile in Pêra-Rio orange juice. Food Research International. 2018;**106**:54-63

[10] Khan MK, Huma ZE, Dangles O. A comprehensive review on flavanones, the major citrus polyphenols. Journal of Food Composition and Analysis. 2014; **33**:85-104

[11] Gil-Izquierdo A, Gil MI, Ferreres F, Tomás-Barberán FA. In vitro availability of flavonoids and other phenolics in orange. Journal of Agricultural and Food Chemistry. 2001;**49**:1035-1041

[12] He X, Lian L, Lin L, Bernart MW. High-performance liquid chromatography-electrospray mass spectrometry in phytochemical analysis of sour orange (Citrus aurantium L.). Journal of Chromatography A. 1997;**791**: 127-134

Chapter 7

HPLC-MS$^{(n)}$ Applications in the Analysis of Anthocyanins in Fruits

Seyit Yuzuak, Qing Ma, Yin Lu and De-Yu Xie

Abstract

Anthocyanins are water-soluble pink/red/blue/purple pigments found abundantly in the flesh and skin of fruits, flowers, and roots of different varieties of plants. Compared to vegetative tissues in many plants, fruits have much higher contents of anthocyanins. In general, anthocyanins have antioxidant, anti-inflammatory, antimutagenic, and antiapoptotic activities that benefit human health. To date, anthocyanins in many different fruits have gained intensive studies in structures, biosynthesis, genetics, and genomics. Despite this, difficulties exist in identifying anthocyanins with similar structures and precisely estimating contents within fruit matrices. To improve this challenge, high-performance liquid chromatography coupled to tandem mass spectrometry (HPLC-MS/MS) based metabolomics has been shown a powerful technology to distinguish structure-similar anthocyanins. This chapter reviews, summarizes, and discusses the application of HPLC-MS/MS in the annotation or identification of anthocyanins in fruits.

Keywords: anthocyanins, chromatography, HPLC, mass spectrometry, fruits

1. Introduction

Anthocyanins are a class of plant flavonoids belonging to polyphenolics. Anthocyanins are water-soluble pigments that give pink/red/purple/blue color to plant tissues. Anthocyanins are found in the majority of higher plant species except in plant species of Caryophyllales. Moreover, anthocyanins have been found in some lower plants, such as mosses and ferns [1]. Certainly, anthocyanins are important agronomical traits in many crops, particularly ornamental ones for flowers and fruits [2]. Anthocyanins are synthesized in the cytosol and mainly transported to the central vacuoles. Plant cells such as epidermal cells in the peel of fruits and flower petals are the main locations with the active biosynthesis of anthocyanins. Generally, anthocyanins are stored as the colored flavylium ion form due to the acidic conditions of the vacuoles [3]. The color changes of plant tissues are normally associated with pH value variations in the central vacuoles.

1.1 Structure

All anthocyanins are derived from a specific chromophore core, namely 2-phenyl-benzopyrylium or flavylium, which consists of two aromatic rings (A and B) and one heterocyclic pyran ring including three carbons (C) (**Figure 1**), thus is featured by C6-C3-C6. Seven positions (R_3, R_5, R_6, R_7, $R_{3'}$, $R_{4'}$, $R_{5'}$) are commonly subjected to modification of monosaccharides, methyl, or other groups *via* –OH or carbon [4, 5]. Past structural studies have shown that anthocyanins predominantly found in nature are glycosylated or galloylated 2-phenyl-benzopyrylium or flavylium salts in acidic conditions. The flavylium structures also are subjected to modification by the vacuolar pH values to give different hue features of plant tissues.

To date, more than 700 distinct anthocyanins have been identified in the plant kingdom. Structurally, anthocyanins are composed of an aglycone also called anthocyanidin and carbohydrate residue. All anthocyanins are derivatives of anthocyanidin aglycones. Although different reports have listed 19 anthocyanidins (6 major and 13 minor groups) (**Table 1**), based on the biosynthesis from phenylalanine and three malonic acids, pelargonidin, cyanidin, and delphinidin form the three basic ones. Other types such as peonidin, petunidin, and malvidin result from methylation of the B-ring of the three basic forms. Actually, methylated anthocyanidins have different physical and chemical features from their precursors. To date, pelargonidin, cyanidin, delphinidin, peonidin, petunidin, and malvidin are commonly accepted to represent six major groups of anthocyanidins [6–8]. The glycosylation or methylation, acylation, galloylation, and other modifications diversify anthocyanin structures. Based on the structures reported, glycosides of anthocyanidin form the most predominant structures in nature. Past investigations have reported that glycosylation mainly occurs at the –OH group of C_3, C_5, or C_7 of the core chromophore. The main monosaccharides involved in glycosylation include glucose, xylose, arabinose, rhamnose, rutinose, fructose, and galactose moieties. In addition, disaccharides are involved in glycosylation. The acylation, acetylation, and malonylation of anthocyanidins or anthocyanins lead to more diversity of structures. Based on structural and biosynthetic reports, the common acylation result from the addition of a coumaric, caffeic, ferulic, *p*-hydroxy benzoic, synaptic, malonic, acetic, succinic, oxalic, or malic acid to sugar moiety or moieties [7, 9]. The acylation can alter the physical or chemical features of anthocyanins, such as water solubility and color such as blue color. In addition, the hydroxyl and methoxyl groups as well as other modifications also lead to different colors and stability of anthocyanins [6, 8]. Furthermore, secondary, ternary, and more complicated modifications on sugars or acylate groups diversify

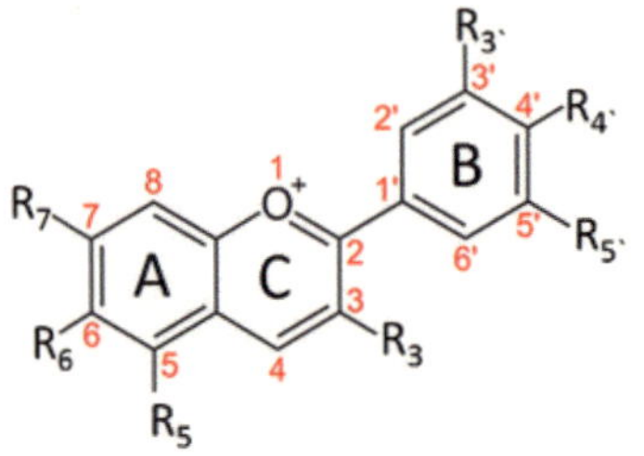

Figure 1.
The core chemical structure of anthocyanidins. A; aromatic ring, B; phenyl ring, C; Benzopyran ring. R; -H, -OH, or -CH_3.

Anthocyanidins	Substitutions						
	R_3	R_5	R_6	R_7	$R_{3'}$	$R_{4'}$	$R_{5'}$
Cyanidin (**Cy**)	OH	OH	H	OH	OH	OH	H
Delphinidin (**Dp**)	OH	OH	H	OH	OH	OH	OH
Pelargonidin (**Pg**)	OH	OH	H	OH	H	OH	H
Peonidin (**Pn**)	OH	OH	H	OH	OMe	OH	H
Petunidin (**Pt**)	OH	OH	H	OH	OMe	OH	OH
Malvidin (**Mv**)	OH	OH	H	OH	OMe	OH	OMe
Hirsutidin	OH	OH	H	OMe	OMe	OH	OMe
Rosinidin	OH	OH	H	OMe	OMe	OH	H
Aurantinidin	OH	OH	OH	OH	H	OH	H
6-Hydroxycyanidin	OH	OH	OH	OH	OH	OH	H
6-Hydroxydelphinidin	OH	OH	OH	OH	OH	OH	OH
Capensinidin	OH	OMe	H	OH	OMe	OH	OMe
Europinidin	OH	OMe	H	OH	OMe	OH	OH
5-Methylcyanidin	OH	OMe	H	OH	OH	OH	H
Pulchellidin	OH	OMe	H	OH	OH	OH	OH
Luteolinidin	H	OH	H	OH	OH	OH	H
Apigenidin	H	OH	H	OH	H	OH	H
Tricetinidin	H	OH	H	OH	OH	OH	OH
Riccionidin A	H	H	OH	OH	H	OH	H

Table 1.
Basic and methylated or hydroxylated anthocyanidins (OMe: Methoxy group; OH: Hydroxyl group; H: Hydrogen) modified from ref. [6].

anthocyanin structures. Most of those modifications result from enzymatic reactions catalyzed by glycosyltransferases, methyltransferases, acyltransferases, malonyltransferases, and others [4, 10]. These modifications are associated with the color, chemical stability, bioavailability, biological activity, and diverse molecular structures of anthocyanins in plants.

1.2 Functions in plants and health benefits

Anthocyanins play important physiological functions associated with plant reproduction and defenses. Anthocyanins act as visual signals to attract pollinators for the pollination of flowers and dispersers to spread seeds or fruits [4]. Anthocyanins can play as warning signals to repulse birds and insects for protection of plant tissues from being consumed [11]. They serve as filters to absorb UV-B light and visible light for protecting plant tissues from being damaged by severe irradiation [12]. The accumulation of anthocyanins protects leaves from radiation-caused damage of photosynthesis by absorbing extra light [13]. Past studies have reported that many anthocyanin structures defend plants against diseases infected by various pathogens and damage caused by abiotic stresses, including cold shock, drought, osmotic and wounding, and biotic stresses [5, 14, 15]. Furthermore, anthocyanidins are essential precursors of proanthocyanidins, which are powerful antioxidative, anti-pathogen, anti-radiative, and anti-pest flavonoids in most of the plant species [16].

Anthocyanins are of growing interest in beneficial values for human and animal health because of their antioxidative, antibacterial, and anticancer activities reported by *in vitro* experiments [17]. Some structures have been shown to scavenge free radicals such as reactive oxygen species generated in human cells [18]. More studies have shown that the antioxidative and anti-inflammation activities of anthocyanins help to improve visual acuity [19] and to protect from a heart attack [20, 21]. Metabolic studies have shown that certain anthocyanin structures can prevent obesity and diabetes by interfering the body weight gain and adipose tissue increase [22, 23]. Especially, different structures have been observed to associate with specific activities [24]. Accordingly, a general health fact is that regular consumption of high levels of dietary anthocyanin-rich plant sources, such as red, purple, and dark-colored berries, grapes, and vegetables, is considered to benefit human health [25].

2. Analysis of anthocyanins in fruits

Anthocyanins are the main pigments responsible for red, magenta, violet, blue, and dark blue colors of many fruits and berries. Therefore, fruit or seed anthocyanins have gained a large number of studies, which include the formation and stability of color, and color changes during ripening, processing and storage, isolation, and identification from fruits (**Table 2**). Many have been developed as colorants for food, pharmaceutical, and nutritional industries.

Fruits are one of the main sources of novel anthocyanins with valuable health benefits [8, 26, 27]. A large number of anthocyanins have been isolated from fruits (**Table 2**).

To date, methods for the identification of fruit anthocyanins have been appropriately established by numerous studies. In generation, three main sequential steps are sample preparation, separation and purification, and identification.

Sample preparation, as an initial step for anthocyanin determination, is highly variable depending on the fruit samples and the objectives of the analysis. Sample preparation mainly consists of collection, drying, powdering, and extraction of samples. Liquid samples of fruit, such as juices, wines, and syrups, need very little preparation before the analysis. However, solid or dried fruit materials require to be fractionated, homogenized, crushed, or pulverized. The most commonly used solvents for the extraction of anthocyanins from fruits are the mixtures of ethanol, methanol, acetonitrile, and acetone compositions of water. Depending on the aim of the research, various types and different compositions of solvents can be implemented. For example, aqueous acetone solvents can be mostly preferred for a higher yield, an efficient and more reproducible extraction of anthocyanin [57, 58]. There is no universal and simplified sample preparation method to extract anthocyanins from fruit samples. However, a variety of modern techniques, including solid-phase extraction (SPE), accelerated solvent extraction (ASE), microwave-assisted extraction (MWE), ultrasound-assisted extraction (UAE), pressurized hot water extraction, and supercritical fluid extraction (SFC), have been developed based on maximizing the highest recovery, minimizing the amount of non-anthocyanins and degradation or alteration of the native anthocyanins [59].

Fruit types (Year)	Sample treatment	Anthocyanin type	HPLC/MS features, stationary and mobile phases	Ref. #
• Cranberry, • Highbush blueberry, • Lowbush blueberry • Tifblue blueberry (2001)	Material: Freeze-dried berries. Extraction: Acetone, water, and acetic acid (70:29.5:0.5, v/v). Sonication for 5 min in a water bath at 50 °C. Dilution with 20% methanol/water (v/v). Purification through Sephadex LH-20 column. Anthocyanin elution with 40 mL of 60% methanol/water (v/v).	Dp-3-galactoside, Dp-3-glucoside, Cy-3-galactoside, Dp-3-arabinoside, Cy-3-glucoside, Pt-3-galactoside, Pt-3-glucoside, Cy-3-arabinoside, Pn-3-galactoside, Pt-3-arabinoside, Mv-3-galactoside, Pn-3-glucoside, Mv-3-glucoside, Pn-3-arabinoside, Mv-3-arabinoside, Dp-6-acetyl-3-glucoside, Cy-6-acetyl-3-glucoside, Mv-6-acetyl-3-galactoside, Pt-6-acetyl-3-glucoside, Pn-6-acetyl-3-glucoside, Mv-6-acetyl-3-glucoside	HPLC/DAD/ESI-MS/MS Positive ion mode. Reverse Phase. Zorbax C18 column (150 x 4.6 mm) Solvent A; 5% formate acid B; 100% methanol. Flow Rate: 1.0 mL/min	[28]
• Evergreen blackberry (2002)	Material: Frozen berries. Powdered by liquid nitrogen. Extraction: Acetone (1:1, w/v). Treatment with chloroform (1:2, v/v). Purification through C18 Sep-Pak cartridges. Anthocyanin elution with methanol containing concentrated formic acid (95:5, v/v).	Cy-3-glucoside, Cy-3-rutinoside, Cy-xylose, Cy-3-glucoside acylated with malonic acid	HPLC/PDA/ESI-MS/MS Positive ion mode. Reverse phase. Prodigy C18 column (250 x 4.6 mm, 5 ųm) Solvent A; 100% acetonitrile B; A mixture of 1% phosphoric acid (85%), 10% acetic acid (glacial), and 5% acetonitrile (v/v/v) in water. Flow Rate: 1.0 mL/min	[29]
• Red raspberry (2002)	Material: Fresh fruits. Macerated. Extraction: Methanol containing 0.1% HCl	Cy-3-sophoroside, Cy-3-(2 -glucosylrutinoside), Cy-3-glucoside, Pg-3-sophoroside, Cy-3-rutinoside, Pg-3-(2 -glucosylrutinoside), Pg-3-glucoside, Pg-3-rutinoside	HPLC/UV-Vis/APCI-MS/MS Positive ion mode. Reverse phase. Novapac C18 column (250 x 4.6 mm, 5 ųm) Solvent A; 100% acetonitrile B; 1% formic acid. Flow Rate: 0.8 mL/min	[30]
• Linden arrowwood (2003)	Material: Squeezed fresh fruit juice. Purification through Sephadex LH-20 column (700mm 25 mm). Anthocyanin elution with H2O, 20% methanol containing 0.1 N HCl acid.	Cy 3-sambubioside, Cy-3-O-,-D-glucopyranoside (namely kuromanin)	HPLC/PDA/ESI-MS/MS Positive ion mode. Reverse phase. Capcell Pak C18 column (150 x 2.0 mm) Solvent A; 5% formic acid B; acetonitrile. Flow Rate: 1.0 mL/min	[31]

Fruit types (Year)	Sample treatment	Anthocyanin type	HPLC/MS features, stationary and mobile phases	Ref. #
• Black, Green, Red and White currants (2003)	Material: Frozen fresh berries. Extraction: Removal of other phenolics by ethyl acetate. Acidification of the residue with hydrochloric acid (2 M, 2 mL), and anthocyanins extraction with methanol.	Dp 3-O-glucoside, Dp 3-O-rutinoside, Cy 3-O-sophoroside, Cy 3-O-(2G-glucosylrutinoside), Cy 3-O-sambubioside, Cy 3-O-glucoside, Cy 3-O-(2G-xylosylrutinoside), Cy 3-O-rutinoside	HPLC/DAD/ESI-MS/MS Positive ion mode. Reverse phase. Purospher RP-18e column (125 x 3 mm, 5 ɥm) Solvent A; 5% formic acid B; acetonitrile. Flow Rate: 0.5 mL/min	[32]
• Bilberry, • Rab-biteye blueberry, • Black currant (2004)	Material: Frozen fruits. Extraction: 450 mL of 90% ethanol (0.1% H2SO4) at RT. Purification through a column of non-ionic polymeric absorbent (Amberlite XAD-7). Anthocyanin elution with aqueous ethanol (0.05% citric acid).	Dp 3-galactoside, Dp 3-glucoside, Dp 3-rutinoside, Cy 3-galactoside, Dp 3-arabinoside, Cy 3-glucoside, Pt 3-galactoside, Cy 3-rutinoside, Cy 3-arabinoside, Pt 3-glucoside, Pt 3-rutinoside, Pn 3-galactoside, Pt 3-arabinoside, Pn 3-glucoside, Mv 3-galactoside, Pn 3-rutinoside, Mv 3-arabinoside	HPLC/PDA/ESI-MS/MS Positive ion mode. Reverse phase. Capcell Pak C18 column (150 x 4.6 mm, 5 ɥm) Solvent A; 0.1% TFA/H_2O B; 50% CH_3CN/0.1% TFA/50% H2O. Flow Rate: 0.5 mL/min	[33]
• Red and White grape cultivars (2004)	Material: Frozen grape pomace. Lyophilized and finely grounded. Extraction: 100 mL of methanol/0.1% HCl (v/v) for 2 h.	Dp 3-O-glucoside, Cy 3-O-glucoside, Pt 3-O-glucoside, Pn 3-O-glucoside, Mv 3-O-glucoside, Dp 3-O-acetylglucoside, Pt 3-O-acetylglucoside, Pn 3-O-acetylglucoside, Mv 3-O-acetylglucoside, Cy 3-O-p-coumaroylglucoside, Pt 3-O-p-coumaroylglucoside, Pn 3-O-p-coumaroylglucoside, Mv 3-O-p-coumaroylglucoside	HPLC/DAD/ESI-MS/MS Positive ion mode. Reverse phase. Aqua C18 column (250 x 4.6 mm, 5 ɥm) Solvent A; water/formicacid/acetonitrile (87:10:3, v/v/v) B; water/formic acid/ acetonitrile (40:10:50, v/v/v). Flow Rate: 0.8 mL/min	[34]
• Raspberry, • Arctic bramble, • Cloudberry, • Strawberry (2004)	Material: Frozen berries. Extraction: Ethyl acetate (410 mL), acidification of the berry residue with HCl (2 M, 2 mL), and anthocyanins extraction as flavylium cations with methanol.	Cy 3-hexoside-deoxyhexoside, Cy 3-sophoroside, Cy 3-(2G-glucosylrutinoside), Cy 3-glucoside, Cy 3-rutinoside, Cy glycoside, Pg 3-glucoside, Pg 3-rutinoside, Pg 3-malonylglucoside, Pg 3-succinylglucoside	HPLC/DAD/ESI-MS/MS Positive ion mode. Reverse phase. Purospher RP-18e column (125 x 3 mm, 5 ɥm) Solvent A; 5% formic acid B; acetonitrile. Flow Rate: 0.5 mL/min	[35]

Fruit types (Year)	Sample treatment	Anthocyanin type	HPLC/MS features, stationary and mobile phases	Ref. #
• Butcher's broom (2005)	Material: Frozen berries. Extraction: 0.1% HCl (v/v) in methanol for 20 h at RT, in the dark. Purification through C-18 Sep-Pak cartridge. Anthocyanin elution with methanol containing 0.01% HCl (v/v).	Pg-3-glucoside, Pg-3-rutinoside, Pg-3-cis-p-coumarylglucoside, Pg-3-trans-p-coumarylglucoside	HPLC/DAD/ESI-MS/MS Positive ion mode. Reverse phase. Wakosil C18 column (150 x 4.6 mm, 5 ɥm) Solvent A; 0.1% TFA in water B; 0.1% TFA in acetonitrile. Flow Rate: 1.0 mL/min	[36]
• Buckthorn (2005)	Material: Frozen berries. Extraction: 100 mL of 0.1% HCl (v/v) in methanol for 20 h at RT in the dark. Purification throug C-18 Sep-Pak cartridge. Anthocyanin elution with methanol containing 0.01% HCl (v/v).	Dp 3-rutinoside, Cy 3-rutinoside, Pt 3-rutinoside, Pg 3-rutinoside, Pn 3-rutinoside, Mv 3-rutinoside	HPLC/DAD/ESI-MS/MS Positive ion mode. Reverse phase. Wakosil C18 column (150 x 4.6 mm, 5 ɥm) Solvent A; 0.1% TFA in water B; 0.1% TFA in acetonitrile. Flow Rate: 1.0 mL/min	[37]
• Red Kiwi cultivars (2005)	Material: Fruit pericarp Extraction: 5 times volume (v/w) of ethanol/H2O/acetic acid (80:20:1 v/v/v) in an Ultra-Turrax homogenizer for 2 days at 4°C in the dark.	Cy 3-O-xylo (1–2)-galactoside, Cy 3-O-galactoside, Cy 3-O-glucoside, Cy 3-O-arabinoside	HPLC/PDA/ESI-MS/MS Positive ion mode. Reverse phase. LiChroCart Superpher 100 RP-18 end-capped column (250 x 2 mm) Solvent A; methanol/formic acid/water (v/v/v, 15:3.75:81.25) B; 100% methanol. Flow Rate: 0.25 mL/min	[38]
• Camu-camu (2005)	Material: Fruit peel. Freeze-dried. Extraction: 1000 mL of 0.5% HCl in methanol overnight at 5°C, in darkness. Purification through an Amberlite XAD-7 resin open column.	Dp-3-glucoside, Cy-3-glucoside	HPLC/Uv-Vis/ESI-MS/MS. Positive ion mode. Reverse Phase. RP-12 Synergi MaxRO column (250 x 4.6 mm, 5 ɥm) Solvent A; water/acetonitrile/formic acid (87:3:10, v/v/v) B; water/acetonitrile/formic acid (40:50:10, v/v/v). Flow Rate: 0.5 mL/min	[39]

Fruit types (Year)	Sample treatment	Anthocyanin type	HPLC/MS features, stationary and mobile phases	Ref. #
• Plum, • Black plum, • Nectarine, • Peach, • Marionberry (2005)	Material: Freeze-dried fruits. Powdered. Extraction: Methanol/water/acetic acid (85:15:0.5, v/v, MeOH/H2O/AcOH).	Cy 3-galactoside, Cy 3-xyloside, Cy 3-glucoside, Cy 3-(6″-acetoyl) glucoside, Cy 3-rutinoside, Cy 3-glucoside, Pg 3-glucoside, Cy 3-(maloyl) glucoside, Pn 3-glucoside, Cy 3-(6″-acetoyl) glucoside, Pn 3-rutinoside, Cy 3-(6″-malonoyl) glucoside, Cy 3-dioxaloylglucoside	HPLC/DAD/ESI-MS/MS Positive ion mode. Reverse Phase. Zorbax Stablebond Analytical SB-C18 column (4.6 x 250 mm, 5 ɥm) Solvent A; aqueous 5% formic acid B; methanol. Flow Rate: 1 mL/min	[40]
• Lowbush blueberries (2006)	Material: Fresh fruits. Extraction by preparative techniques (data not shown).	Dp 3-O-galactoside, Dp 3-O-glucoside, Dp 3–O-arabinoside, Cy 3-O-galactoside, Cy 3-O-glucoside, Cy 3-O-arabinoside, Pt 3-O-galactoside, Pt 3-O-glucoside, Pt 3-O-arabinoside, Pn 3-O-galactoside, Pn 3-O-glucoside, Mv 3-O-galactoside, Mv 3-O-glucoside, Mv 3-O-arabinoside	HPLC/PDA/ESI-MS/MS Positive ion mode. Reverse Phase. Zorbax SB-C8 column (150 x 62.1 mm, 3.5 ɥm) Solvent A; aqueous 10% formic acid B; Acetonitrile. Flow Rate; 0.35 mL/min	[41]
• Wild blueberries (2007)	Material: Frozen berries. Grounded and homogenized. Extraction: Ethanol at 7.7°C, 26°C, or 79°C without acid (pH 5.4) or acidified with HCl (pH 4.1), citric acid (pH 4.9), tartaric acid (pH 5.0), lactic acid (pH 4.8), or phosphoric acid (pH 4.6; 0.02%, v/v).	Dp 3-galactoside, Dp 3-glucoside, Dp 3-hexose, Cy 3-galactoside, Pt 3-galactoside, Mv 3-galactoside, Dp 3-arabinoside, Pt 3- glucoside, Cy 3-glucoside, Mv 3- glucoside, Pt hexose, Cy 3-arabinoside, Pn 3-galactoside, Mv 3-hexose, Pn 3-glucoside, Mv hexose, Pt propionyl-hexose, Cy malonyl-hexose, Mv 3-arabinose, Dp -oxalyl-hexose, Cy 3-propionyl-galactoside, Cy -malonyl-hexose, Cy 3-malonyl-galactoside, Pt -malonyl-pentose, Cy malyl-pentose, Dp succinyl-pentose, Pt -propionyl-hexose, Pt 3-propionyl-glucoside/-hexose, Pt 3-acetyl-glucoside, Pn 3-succinyl-arabinoside, Pn 3-oxalyl-galactoside, Mv 3-acetyl-hexose, Mv oxalyl-pentose, Mv 3-acetyl-glucoside, Mv 3-acetyl-galactoside	HPLC/DAD/ESI-MS/MS Positive ion mode. Reverse Phase. Zorbax Stablebond Analytical SB-C18 column (4.6 x 250 mm, 5 ɥm) Solvent A; aqueous 5% formic acid B; methanol. Flow Rate; 1 mL/min	[42]

Fruit types (Year)	Sample treatment	Anthocyanin type	HPLC/MS features, stationary and mobile phases	Ref. #
• Lycium ruthenicum fruit (2015)	Material: Fresh fruits. Extraction: 70% ethanol for 2 h at pH 2.5 (adjusted with HCl). Extraction: Purification through AB-8 macroporous resin. Anthocyanin elution with 70% ethanol.	Pt-3-O-rutinoside-5-O-glucoside, Mv-3-O-rutinoside-5-O-glucoside, Dp-3-O-rutinoside(glucosyl-trans-p-coumaroyl)-5-O-glucoside, Pt-3-O-rutinoside(glucosyl-trans-p-coumaroyl)-5-O-glucoside, Pt-3-O-rutinoside(glucosyl-cis-p-coumaroyl)-5-O-glucoside, Mv-3-O-rutinoside(glucosyl-trans-p-coumaroyl)-5-O-glucoside, Dp-3-O-rutinoside(cis-p-coumaroyl)-5-O-glucoside, Pt-3-O-rutinoside(trans-caffeoyl)-5-O-glucoside, Mv-3-O-rutinoside(glucosyl-cis-p-coumaroyl)-5-O-glucoside, Dp-3-O-rutinoside(trans-p-coumaroyl)-5-O-glucoside, Pt-3-O-rutinoside(cis-p-coumaroyl)-5-O-glucoside, Pt-3-O-rutinoside(trans-p-coumaroyl)-5-O-glucoside, Pt-3-O-rutinoside(feruloyl)-5-O-glucoside, Mv-3-O-rutinoside(cis-p-coumaroyl)-5-O-glucoside, Mv-3-O-rutinoside(trans-p-coumaroyl)-5-O-glucoside, Malvidin-3-rutinoside(feruloyl)-5-O-glucoside	HPLC/DAD/ESI-QTOF-MS/MS Positive ion mode. Reverse Phase. XUnion C8 column (4.6 x 150 mm, 5 ɥm) Solvent A; acetonitrile B; 0.5% (v/v) formic acid/water. Flow Rate; 0.2 mL/min	[43]
• Indian blackberry (2015)	Material: Fresh fruit. Pulp filtration through a sieve. Extraction: Acidified (0.05% HCl) aqueous methanol (80% methanol) using an ultra sonicator. Liquid-liquid partitioning with n-hexane, chloroform, ethyl acetate, n-butanol and aqueous methanol.	Cy-3-glucoside, Cy-3,5-diglucoside, Cy-3-diglucoside-5-glucoside, Cy-3-(p-hydroxybenzoyl) (oxaloyl) diglucoside-5-glucoside, Cy-3-(sinapoyl) glucoside-5-glucoside, Dp-3-glucoside, Dp-3,5-diglucoside, Dp acetyl-diglucoside, Mv-3-glucoside, Mv-3,5-diglucoside, Mv-3-(6″-acetoyl)glucoside, Mv-3-(6″-acetoyl) glucoside-5-glucoside, Pn-3-glucoside, Pn-3,5-diglucoside, Pt-3-glucoside, Pt-3,5-diglucoside	HPLC/PDA/MALDI-ESI-MS/MS Positive ion mode. Reverse Phase. Phenomenx C18 colum (4.6 x 250 mm, 5 ɥm) Solvent A; 5% HCOOH in water B; 100% acetonitrile. Flow Rate; 1 mL/min	[44]

Fruit types (Year)	Sample treatment	Anthocyanin type	HPLC/MS features, stationary and mobile phases	Ref. #
• Highbush blueberries, • Murta, • Calafate, • Arrayán, • Chequén, • Meli (2015)	Material: Fresh fruits. Homogenized in a blender and freeze-dried and pulverized. Defatted thrice with 100 ml of n-hexane. Extraction: 100 ml of 0.1% HCl in MeOH in the dark in an ultra-sonic bath for one hour. Purification through XAD-7 (100 g) column. Anthocyanin elution with 100 ml of MeOH acidified with 0.1% HCl.	Pt-3-O-di-hexoside, Cy-3-O-di-hexoside, Dp 3-O-galactoside, Dp 3-O-glucoside, Cy 3-O-rutinose, Cy-3-O-galactoside, Cy-3-O-glucoside, Pt-3-O-rutinoside, Mv-3-O-rutinose, Pt-3-O-glucoside, Pt-3-O-galactoside, Pn-3-O-rutinose, Pn-3-O-galactoside, Mv-3-O-galactoside, Dp-3-O-arabinoside, Pn-3-O-glucoside, Mv-3-O-glucoside, Pn-3-O-arabinoside, Pt-3-O-arabinoside, Cy-3-O-arabinoside, Mv 3-O-(6″coumaroyl) glucoside, Mv-3-O-arabinose, Cy-3-O-(6″succinoyl)-glucose, Pn 3-O-di hexoside, Dp-3-O-rutinose (6″-p-coumaroyl)-2″-O-glucose, Dp 3-O-(6″acetyl) glucoside, Cy 3-O-(6″acetyl) glucoside, Pt 3-O-(6″acetyl) glucoside, Mv 3-O-(6″acetyl) galactoside, Dp-3-O-(6″caffeoyl)-glucose, Mv 3-O-(6″acetyl) glucoside	HPLC/PDA/ESI-TOF-MS/MS Positive ion mode. Reverse Phase. Luna C-18 column (4.6 x 250 mm, 5 ɥm) Solvent A; 0.1% aqueous formic acid B; acetonitrile 0.1% formic acid. Flow Rate; 0.5 mL/min	[45]
• Pomegranate cultivars (2016)	Material: Fresh fruits. Pressurized to make juice. Extraction: 5 mL of MeOH/water (80:20 v/v) containing 2 mM NaF and then centrifuged at 15,000g for 15 min.	Dp 3,5-O-diglucoside, Cy 3,5-O-diglucoside, Dp 3-O-diglucoside, Cy 3-O-diglucoside, Pg 3-O-diglucoside	HPLC/DAD/ESI-MS/MS Positive ion mode. Reverse Phase. XRs 5 C18 column (250 x 4.6 mm, 5 ɥm) Solvent A; water-formic acid (95:5, v/v) B; acetonitrile. Flow Rate; 0.8 mL/min.	[46]
• Chokeberry (2016)	Material: Fresh fruits and pomace. Powdered. Extraction: 50 mL of methanol acidified with 2.0% formic acid. Sonication for 20 min with occasional shaking. Centrifugation at 19,000×g for 10 min.	Cy-3-hexoside-(epi) catechin, Cy-3-pentoside-(epi) catechin, Cy-3-hexoside-(epi) cat-(epi) ca, Cy-3-O-galactoside, Cy-3-O-glucoside, Cy-3-O-arabinoside, Cy-3-O-xyloside	UHPLC/PDA/ESI-MS/MS Positive ion mode. Reverse Phase. BEH C18 column (2.1 x 100 mm, 1.7 ɥm) Solvent A; 2.0% formic acid, v/v) B; 100% acetonitrile. Flow Rate; 0.45 mL/min	[47]

Fruit types (Year)	Sample treatment	Anthocyanin type	HPLC/MS features, stationary and mobile phases	Ref. #
• Myrtus Berries (2017)	Material: Fresh berries. Extraction: 50 mL of 0.1% HCl (v/v) in methanol for 24 h in the dark.	Dp 3-O-glucoside, Pt 3-O-glucoside, Mv 3-O-glucoside Dp-pentose, Dp-pentose, Cy 3-O-glucoside, Pt-pentose. Pn 3-O-glucoside	HPLC/DAD/ESI-QTOF-MS/MS Positive ion mode. Reverse Phase. Gemini-NX C18 column (250 x 4.6 mm, 5 ɥm) Solvent A; 0.1% formic acid in water B; 0.1% formic acid in acetonitrile. Flow Rate; 0.3 mL/min	[48]
• Mulberry (2017)	Material: Fresh fruits. Extraction: Acetone-water-acetic acid (75:14.9:0.1, v/v/v), through homogenization (900Kr/min, 10min) and centrifugation (10,000 rpm, 10min). Purification by a polyamide resin column (4×100 cm). Anthocyanin elution with 10%, 30%, 50%, or 80% (v/v) ethanol solutions containing 0.1% HCl.	Cy 3-sophoroside, Cy 3,5-diglucoside, Pn 3-rutinoside Cy 3-laminaribioside, Pt 3-arabinoside, Dp 3-rutinoside, Cy 3-O-(diglucoside)-glucosylrutinoside, Dp 3-galactoside, Cy 3-glucoside, Cy 3-rutinoside, Dp 3,5-diglucoside, Dp 3-rutinoside-5-glucoside, Cy 3-(glucosyl rhamnoside, Pg 3-glucoside, Cy 3-galactoside, Pg 3-rutinoside	UHPLC/DAD/HR-ESI-TOF-MS/MS Positive ion mode. Reverse Phase. ZORBAX SB-C18 column (2.1 x 50 mm, 1.8 ɥm) Solvent A; 0.1% (v/v) formic acid in water B; acetonitrile. Flow Rate; 0.2 mL/min	[49]
• Nitraria tangutorun fruit (2017)	Material: Dried fruits. Grounded into powder with liquid nitrogen in a grinder Extraction: 15 mL of aqueous ethanol (0.1% HCl, v/v). Purification through a column (16 mm × 50 mm) contained X-5 resin (Bed volume of 60 mL). Anthocyanin elution with 120 mL of ethanol–water–HCl (80:19:1, v/v/v; pH 1) at a flow rate of 2 mL/min for 1 h.	Mv-3-O-diglucoside-5-O-arabinose, Cy-3-O-diglucoside, Cy-3-O-galactoside, Pg-3-O-diglucoside, Pn-3-O-diglucoside, Cy-3-O-(caffeoyl)-diglucoside, Dp-3-O-(p-coumaroyl)-diglucoside, Cy-3-O-(cis-p-coumaroyl)-diglucoside, Cy-3-O-(feruloyl)-diglucoside, Cy-3-O-(trans-p-coumaroyl)-diglucoside, Pg-3-O-(p-coumaroyl)-diglucoside, Dp-3-O-(p-coumaroyl)-hexose, Cy-3-O-(p-coumaroyl)-diglucoside, Cy-3-O-(p-coumaroyl)-hexose	HPLC/DAD/ESI-MS/MS Positive ion mode. Reverse Phase. Hedera ODS-2 column (150 x 2.1 mm, 3 μm) Solvent A; aqueous formic acid (0.1%, v/v) B; acetonitrile. Flow Rate; 0.2 mL/min	[50]

Fruit types (Year)	Sample treatment	Anthocyanin type	HPLC/MS features, stationary and mobile phases	Ref. #
• Muscadine grape (2017)	Material: Freeze-dried berries. Grounded into powder. Extraction: 50 mL 80:20 (v/v) methanol–water solution containing 0.1 mL/L HCl. Shaking of the extract in water-bath at 30 C in the dark for 16 h. Sonication for 15 min, and centrifugation at 2,000g for 10 min. Purification through C18 Solid phase extraction cartridge. Anthocyanin elution with 10 mL methanol.	Dp-3,5-diglucoside, Cy-3,5-diglucoside, Pt-3,5-diclucoside, Mv-3,5-diglucoside, Pn-3,5-diglucoside	HPLC/DAD/ESI-MS/MS Positive ion mode. Reverse Phase. Ultrasphere Cl8 column (250 x 4.6 mm) Solvent A; O-phosphoric acid/methanol/water (5:10:85, v/v/v) B; acetonitrile. Flow Rate; 0.5 mL/min	[51]
• Chinese dwarf cherry (2019)	Material: Freeze-dried berries. Pulverized in liquid nitrogen. Extraction: 2 mL of 2% formic acid methanol (v/v). Sonication for 20 min, and centrifugation at 10,000g for 10 min.	Cy 3-O-glucoside, Cy 3-O-rhamnosyl-hexoside, Pg 3-O-hexoside, Pg 3-O-rhamnosyl-hexoside, Cy 3-O-acetyl-hexoside, Pg 3-O-acetyl-hexoside	HPLC/PDA/ESI-QTOF-MS/MS Positive ion mode. Reverse Phase. ODS-80Ts QA C18 column (250 x 4.6 mm) Solvent A; 10% formic acid in water (v/v) B; 1% formic acid in acetonitrile (v/v). Flow Rate; 0.8 mL/min	[52]
• Red-fleshed apple fruit (2020)	Material: Fresh fruits. Grounded into a fine powder in liquid nitrogen. Extraction: 60 mL chilled 80% methanol (1:6, w/v) at 4°C for 4 hr.	Cy-3-galactoside, Cy-3-arabinoside	UHPLC/ESI-MS/MS Positive ion mode. Reverse Phase. Acquity BEH C18 column (100 x 2.1 mm, 1.7 μm) Solvent A; 0.1% formic acid (v/v) B; acetonitrile. Flow Rate; 0.3 mL/min	[53]
• Goji berry (2021)	Material: Frozen and homogenized fruits. Treatment with solid NaF (2 mM) to inactivate polyphenol oxidases and prevent phenolic degradation. Extraction: Methanol, acetone or water with hydrochloric, acetic, citric and ascorbic acid for 30 min, at 25 °C. Sonication for 30 min. Centrifugation for 15 min at 3000 rpm at RT.	Pt-3-O-(glucosyl-trans-p- coumaroyl)-rutinoside-5-O-glucoside, Pt-3-O-(glucosyl-cis-p-coumaroyl)-rutinoside-5-O-glucoside, Pt-3-O-(caffeoyl)-rutinoside-5-O-glucoside, Pt-3-O-(trans-p-coumaroyl)-rutinoside-5-O-glucoside,Mv-3-O-(p-coumaroyl)-rutinoside-5-O-glucoside, Pt-3-O-(p-coumaroyl)-rutinoside	HPLC/DAD/ESI-MS/MS Positive ion mode. Reverse Phase. Supelco C18 column (250 x 4.6 mm, 5 μm) Solvent A; water-formic acid (2%, v/v) B; acetonitrile. Flow Rate; 0.4 mL/min	[54]

Fruit types (Year)	Sample treatment	Anthocyanin type	HPLC/MS features, stationary and mobile phases	Ref. #
• Northern highbush blueberry cultivars (2022)	Material: Frozen fresh berries. Grounded in liquid nitrogen. Extraction: 1.25 mL of 80 % methanol containing 0.1 % of formic acid (v/v) for 15 min at RT. Centrifugation for 10 min at 3500 rpm, RT. The pellet re-extracted by shaking with 1.25 mL of 65 % acetone containing 0.1 % of formic acid (v/v) for 30 min at RT. Purification: Removal of sugars and proteins by solid-phase extraction (SPE). Anthocyanin elution with 2.5 mL of water containing 2 % formic acid (v/v), and 2.5 mL of methanol containing 0.1 % formic acid (v/v).	Dp-3-O-galactoside, Dp-3-O- glucoside, Cy-3-O-glucoside, Pt-3-O-glucoside, Pn-3-O-glucoside, Mv-3-O-galactoside, Mv-3-O-glucoside	HPLC/DAD/ESI-QTOF-MS/MS Positive ion mode. Reverse Phase. Zorbax SB-C18 column (50 x 4.6 mm) Solvent A; water with 2 % formic acid (v/v), B; acetonitrile with 0.1 % formic acid (v/ v)	[55]
• Mulberry (2022)	Material: Fresh berries. Freeze-dried and fine powdered. Extraction: acidified 95% ethanol (0.1% trifluoroacetic acid v/v). The concentrated extract was dissolved in the acidified methanol (7% acetic acid v/v) to stabilize the anthocyanins, converting anthocyanins to flavylium cations. Purification: 001X7 cation-exchange column. Anthcoyanin elution with a mixture of methanol/NaCl solution (1 M) 1:1 (v/v)	Cy-3-*O*-glucoside and Cy-3-*O*-rutinoside	HPLC-/DAD/ESI-MS Positive ion mode. Reverse Phase. Acclaim TM C18 column (4.6 mm × 250 mm, 5 μm) Solvent A; formic acid and water (1:100 v/v), B; acetonitrile. Flow rate; 0.2 mL/min	[56]

Table 2.
Fruit anthocyanin examples annotated or identified by HPLC-MS/MS-based profiling. Cy, cyanidine; Dp, delphinidin; Pg, pelargonidin; mv, malvidin; Pn, peonidin; Pt, petunidin.

Because of the structural diversity of anthocyanins and their instability in light, high temperature and other conditions, it is necessary to avoid degradation during the sample preparation. In aqueous solutions, anthocyanins exist in four major forms, including the red flavylium cation, the blue quinonoidal base, the colorless carbinol pseudobase, and the colorless chalcone depending on pH. At pH below 2, anthocyanins are found primarily in the form of the red flavylium cation. Hydration of the flavylium cation gives the colorless carbinol pseudobase at pH values from 3 to 6, and the colorless chalcone pseudobase at pH values higher than 6. Since the flavylium cation form of anthocyanins is stable in a highly acidic medium, extraction solvents are required to be acidified by acetic, formic, hydrochloric, or sulfuric acids to prevent the degradation of the non-acylated anthocyanin pigments. However, the excessive acidic condition may have various effects during the extractions such as degradation or partial hydrolysis of the acylated anthocyanins as well. Solvents acidified with 0.1% hydrochloric acid (HCl, v/v) is the most commonly used for the extraction of anthocyanins from fruit samples [57, 58]. Anthocyanins can undergo a structural transformation during the sample treatment under a high temperature. When the temperature is increased, anthocyanins can turn into unstable chalcone formation, and even further degrade to brown products, giving a reduction in the concentration of the major anthocyanins. Therefore, using a lower temperature can improve the stability of anthocyanins during the sample treatment [58]. Temperature values between 30 and 50°C have been mostly used for the extraction of anthocyanins from fruit samples. Light is also destructive to anthocyanins. To minimize the degradation of anthocyanins, it is advisable to perform sample treatment and storage of extracts in darkness [58, 60]. An investigation provided parameters of the sample preparation based on the objectives of the study [61].

Due to the diversity and complexity of plant secondary metabolites, extractions methods for anthocyanins are nonselective and result in solutions with a lot of undesirable substances, such as sugars, proteins, fats, acids, and other water-soluble compounds. Based on this, a further efficient purification and separation method is normally required to remove other substances and elute anthocyanins from the extracts. These processes can normally lead to the loss of minor anthocyanin components, which result from heat, pH, metal complexes, and copigmentation [8, 62]. Undesired components, such as sugars, acids, and other water-soluble compounds, in the crude fruit extracts have been removed with C18 solid-phase extraction (SPE) cartridges containing octadecyl silica and Sephadex LH-20 containing cross-linked dextran resin. In order to purify anthocyanins by adsorption, silicone gels, such as Amberlite IRC 80, Amberlite IR 120, DOWEX50WX8, Amberlite XAD-4, and Amberlite XAD-7HP, have been used. Among the silicone gels, the Amberlite XAD-7HP has been proven to be the most effective resin for anthocyanin purification. Furthermore, other less polar polyphenolics or nonpolar compounds can be removed from the extracts by washing with ethyl acetate, chloroform, butanol, or acetonitrile in acidic conditions. After removal of other undesirable substances, anthocyanin fraction can be eluted with organic solvents, acidified with formic or hydrochloric acids (0.1%, v/v), containing water, ethanol, methanol or their composition in different ratios [57, 59, 63].

Separation of anthocyanins can be carried out by different chromatography techniques that have been developed in the past. Common methods include column chromatography, (CC), counter-current chromatography (CCC), paper chromatography (PC), thin-layer chromatography (TLC), capillary electrophoresis (CE), and high-performance liquid chromatography (HPLC) [61, 62].

A series of methods have also been developed for the characterization of anthocyanin structures. In brief, these methods include UV-visible spectrophotometry (UV-vis), HPLC facilitated with diode array detection or UV-detection, mass spectrometry (MS), tandem mass spectrometry (MS/MS), and nuclear magnetic resonance (NMR) [64, 65]. Since anthocyanins have a specific absorption in the visible wavelengths from 515 to 540 nm, spectrometry has been the main approach for quantification. Unfortunately, this is a measurement of total anthocyanins because it cannot measure specific components in crude extracts [26]. MS and MS/MS technologies are powerful in fragmenting anthocyanin molecules to generate featured fingerprints, which allow the annotation of unknown or known structures. NMR is a powerful technology for assigning hydrogen and carbon. There are different NMR tests such as homo and heteronuclear 2D and 3D techniques [66, 67].

2.1 Chromatographic separation of anthocyanins

Chromatography is a separation technique for the isolation of different compounds in a particular matrix. This type of technique is composed of a stationary phase formed from different materials, and a mobile phase (solvent). The separation of compounds of interest is based on their affinities for the stationary phase. The stationary phase retains the desired compounds, while the mobile phase elutes or migrates undesired substances. Based on the stationary phases, common chromatography techniques include column, paper, and thin-layer chromatography. Based on the mobile phase, the most used technique is liquid chromatography utilizing the physical and chemical features of analytes [68].

2.1.1 Column chromatography (CC) and counter-current chromatography (CCC)

Column chromatography (CC) is an effective method to fractionate and purify anthocyanins. This separation method is based on the different distribution coefficients of anthocyanins in solid and mobile phases. Common materials used for the solid phase packed in the column include macroporous, polyamide, and sephadex resins, which do not contain ion exchange groups. Macroporous resins (MRs) have multiple advantages, such as fast, stronger, and large capacity of adsorption and desorption potential for anthocyanin purification. MRs are useful for the first step of isolation to obtain fractions. Polyamide and sephadex resins are normally used to separate anthocyanins. To date, although the CC is a favorable method for anthocyanin purification in the laboratory, it is a challenge for scale-up purification [62, 69].

Counter-current chromatography (CCC) has an industry-scale technology for the separation and purification of bioactive anthocyanins from a large amount of plant materials. CC is a support-free liquid-liquid chromatography. Its development is based on the fractionation of compounds between immiscible stationary and mobile liquid phases of a biphasic solvent system. In the separation of active anthocyanins, important factors that need to be considered are acidic solvents in the absence of oxygen, linear elution or gradient elution, pH zone refining, and strong ion exchange. The main drawback is that certain organic solvents used are toxic to human health [63, 70].

2.1.2 Paper chromatography (PC) and thin layer chromatography (TLC)

Paper and thin layer chromatography are two simple techniques used for the separation of anthocyanins. PC was one of the earliest methods. It depends on specific samples, different mobile phases, and papers used. The advantage of PC is simple and fast to examine anthocyanins. The disadvantage of PC is the limited capacity to separate scale-up samples and the separation of metabolites is relatively poor. TLC uses silica or cellulose gel or both. The separation capacity depends on silica and cellulose size as well as developing solvents. In comparison, TLC can overcome poor separation problems occurred in PC. To date, these two methods are used in common for anthocyanin research in the laboratory because of simple, fast, and low-cost advantages [6, 65].

2.1.3 Capillary electrophoresis (CE)

Analysis of anthocyanins is difficult because they can undergo structural degradation under alkaline pH, light, and high temperature. Therefore, it is required to perform analytical methods as rapidly as possible for preventing their degradation and to avoid using more solvents during analysis. For this reason, CE is a promising separation technique due to being more rapid than other techniques and using only minor amounts of solvents [71]. CE is a separation method based on the electrophoretic motility of metabolites. This technique has excellent mass sensitivity, high resolution, low sample requirement, and decreased solvent waste. When a sample is introduced into the capillary tube at the anode, the basic or acidic mobile phase migrates some components of the sample toward the cathode while others are stuck at the anode. Because anthocyanins are not stable in basic solvents, acidic solvents are used to maintain protonated the flavylium cation form. In addition, CE is configured from cathode to anode. Based on its charge-to-size ratio, particular anthocyanins or other compounds are migrated in the CE system. The migration time of compounds with higher charge-to-size ratio takes a longer time. Detection of compounds is achieved by the UV-vis spectrophotometry coupled to CE, which records the spectra from 200 to 599 nm for each peak [72–74].

The separation of anthocyanins by the CE method includes the use of fused-silica capillaries non-coated or coated with Bare or poly-LA 313 (polycationic amine-containing polymer). Non-coated capillaries are rarely used because anthocyanins cannot be excluded due to interactions between negatively charged silica surfaces and positively charged anthocyanins. Therefore, the coated capillaries are the most suitable for anthocyanin separation. Several background electrolyte (BGE) buffers, such as ammonium formate, ammonium acetate, borate, acetic acid, formic acid, and mixtures of formic acid and hydrochloric acid, are applied both basic and acidic. In order to get highly efficient separations, MS-compatible volatile BGE buffers are used. Non-volatile borate or phosphate buffers can be also applied but these buffers are not compatible enough with MS. Furthermore, the alkaline pH of borate buffers could cause the degradation of anthocyanins. However, an acidic BGE buffer helps to prevent anthocyanin degradation [71, 75].

2.1.4 High-performance liquid chromatography (HPLC)

High-Performance Liquid Chromatography (HPLC) is a separation technique where the mobile phase is pressurized so that it can flow through the column much more efficiently. HPLC is the most convenient for components that cannot withstand high temperatures. Thus, HPLC is widely used for the qualitative and quantitative analysis of anthocyanins. There are two main types of columns of HPLC depending on the aim of the study, including analytical columns for analysis and preparative columns for isolating and refining specific compounds from samples. HPLC columns are packed with inert materials to form the stationary phase and vary in length and internal diameter. Analytical and preparative columns are normally designed for microgram-scale and milligram-to-gram-scale separation of compounds, and adjusted to the characteristics of each analyte. HPLC utilizes different separation modes depending on the primary characteristics of compounds such as polarity and electrical charge [6, 68, 76].

Depending on polarity, there are two types of HPLC separation modes, normal-phase mode (NP) and reverse-phase mode (RP). The basic principle of these separation modes is that compounds with similar polarity will show much more attraction to each other. Furthermore, the separation result and accuracy will be depending on the retention time and the speed flow rate, respectively. The normal phase system consists of the non-polar mobile phase and polar stationary phase. When the sample enters the column, the metabolites (polar) with similar polarity to the stationary phase are retained, resulting in longer retention time while other metabolites (non-polar) with similar polarity to the mobile phase move along the column with shorter retention time. The retention time differences allow appropriate separation of anthocyanins and other metabolites. Past studies have reported that the NP system is effective only for the separation of proanthocyanidins but not for anthocyanins due to retaining in the NP's polar stationary phase. By contrast, the reverse-phase system uses a polar mobile phase and nonpolar stationary phase. Therefore, RP chromatography is effective for the separation of anthocyanins due to having a similar polarity with the mobile phase. In RP chromatography, compounds with higher polarity elutes earlier than non-polar compounds. For flavonoid analysis, there are different RP separation phases using C8, C12, C18, phenyl or phenyl-hexyl, pentafluorophenyl (PFP), and polar and polymeric embedded RP columns. In general, C8 and C18 columns, which are filled with particles of silica bonded with alkyl chains, have been used to separate anthocyanins in RP chromatography. The majority of anthocyanin separations in fruits have been performed using C18 columns with the column particle size mainly ranging from 1.7 μm to 5.0 μm [68, 76].

Multiple organic solvents have been used as mobile phases to elute anthocyanins from the columns. Commonly used ones include methanol, acetonitrile, isopropanol, or ethanol, which are mixed with water supplemented with acetic acid, formic acid, ammonium acetate, or trifluoroacetic acid to form aqueous/organic elution solvents. In many separation experiments, organic solvents and acidic water are used to develop a gradient binary solvent system for the ideal separation of different structures. Furthermore, the acidic water or solvents allow for maintaining the flavylium cationic species, thus increasing chromatographic performances [61, 76].

RP chromatography is effective for the separation of anthocyanins and anthocyanidin aglycones. This is associated with the solubility of the compounds in the mobile phase solvents. In general, an optimized gradient solvent system such as acetonitrile-water or methanol-water solutions supplemented with 0.11% acetic acid or formic acid are appropriate to elute anthocyanins. The elution order of anthocyanins through RP chromatography is normally a function of the number of hydroxyl groups and their degree of methoxylation. A general rule is that diglycosylated or more glycosylated anthocyanins are eluted earlier in the column, followed by monoglycosylated anthocyanins (in an order of galactosides, glucosides, arabinosides, xylosides, and rhamnosides) and aglycones [76]. For example, the elution order of anthocyanins in grapes follows a trend; delphinidin < cyanidin < pelargonidin < peonidin < malvidin, along with the number of glucosides and their acylation pattern following the order; diglucosylated < monoglucosylated < monoglucosylated–acetic acid < diglucosylated-coumaric acid < monoglucosylated-caffeic acid < monoglucosylated-coumaric acid [57].

Another separation mode being applied in the HPLC is hydrophilic interaction in liquid chromatography (HILIC). The development of this technique is based on polarity and hydrophilicity. HILIC uses a combination of NP's polar stationary phase, RP's aqueous-organic mobile phase, and the net surface charge of compounds (ion exchange), thus is a better separating method for small polar compounds. In HILIC, stationary phases usually consist of polymer-based, bare silica, or modified silica gels (Accucore HILIC and Acclaim HILIC 10). For the mobile phase, high organic water-miscible polar organic solvents and acetonitrile are used, giving a better polar component separation and an optimal retention time. When samples enter the column and move along the stationary phase, an interaction between water and silica will occur while acetonitrile will form layers above, giving a gradient of mobile phase, and retention caused by partitioning. More hydrophilic molecules retain more in the stationary phase. The preference of HILIC for polar metabolites has allowed its application to separate highly polar anthocyanins that cannot be separated by RP. To date, a column packed with ethylene bridged hybrid (BEH) amide (2.5 μm particle size) has been shown to provide efficient separation of diverse structures of glycosylated and acylated anthocyanins in fruits. However, HILIC may not be appropriate for the separation of isomeric anthocyanidin-hexosides and cis/trans acylated anthocyanin isomers [61, 68, 77, 78].

The polarity of anthocyanins can be reduced by acylation occurring at –OH or carbon on A, B, and C rings. Common acylation groups include malic, acetic, malonic, succinic, gallic, protocatechuic, hydroxybenzoic, vanillic, caffeic, syringic, p-coumaric, ferulic, and synaptic acids. Acylated anthocyanins are usually eluted later than glycosylated ones. In addition, the anthocyanidin elution order of RP mode follows this trend; delphinidin < cyanidin < petunidin < pelargonidin < peonidin < malvidin. This is in contrast with the elution order reported from HILIC (hydrophilic interaction in liquid chromatography); malvidin, followed by peonidin, petunidin, cyanidin, and lastly delphinidin. Therefore, a combination of HILIC and RP-LC separation modes is useful for the comprehensive 2-dimensional liquid chromatographic (LC × LC) analysis of anthocyanins [76, 78, 79].

Because of the structural complexity of anthocyanin content, the polarities of different anthocyanin subgroups may yield peaks overlapped, causing unresolved chromatographic peaks. For instance, for a given aglycone base, the molecular masses for the 3,5-diglucoside and the caffeoyl glucoside are identical, resulting in a limitation for precise identification. Furthermore, anthocyanins are not easy to effectively separate from copigments such as phenolic acid and flavanols due to their similar structure as well. To alleviate these limitations associated with conventional C18 reversed-phase methods, an ion-exchange mode has been applied for anthocyanin separation. The stationary phase of ion-exchange columns consists of anion-exchange (AV-17-8, AV-17-2P, and EDE-10P) and cation-exchange (KU-2-8, Primesep B2, SCX, and 001X7) resins, which are having polar fixed groups. While the anion-exchange resins are positively charged, the cation-exchange resins have a negative charge. Depending on the net surface charge of analyte's, the cation-exchange and anion-exchange stationary phases bind with negatively and positively charged compounds, respectively. In the case of the anion-exchange resins, the adsorption capacity increases, while pH of external solutions raises from acidic to neutral and alkaline values. However, anthocyanins can undergo partial degradation and lose their biological functions in alkaline solutions. Therefore, to ensure maximum adsorption and separation, cation-exchange resins are mostly preferred for anthocyanin separation. Depending on the pH of environment, anthocyanins alternate between the cationic flavylium ion and the neutrally charged carbinol or quinoidal forms. In highly acidic conditions, anthocyanins would convert to positively charged flavylium cations because of the hydroxyl group in the 3-position, resulting in retarding on the negatively charged cation-exchange resins through the ionic interaction, and flushing other phenolic compounds that are not likely adsorbed by the resin. The mobile phase of ion-exchange mode is a solution that has counterions in general [56, 80, 81].

The separation mechanism for the ion-exchange is a mode based on the net charge contained in samples and their pH. It starts with an application of the counterion mobile phase including ion charges (Na^+ and Cl^-). After loading of the sample, molecules with different net charges from the stationary phase bind to the resin while other unbound molecules are washed out by increasing the concentration of counterions or pH value of the mobile phase. For example, a cation exchange chromatography with 001X7 resin has been developed for copigments-free anthocyanins isolation both on a small and large scale from mulberry extracts. In this study, acidifed anthocyanin fractions were eluted with a mixture of methanol/NaCl solution. Cation-exchange 001X7 resin has been reported to be more advantageous with more than 95% purity compared to the macroporous adsorbent and strong cation exchange resins for the purification of anthocyanins [56, 81].

A novel separation mode combining both ion-exchange and reversed-phase separation mechanisms has been also developed, called mixed-mode ion-exchange reversed-phase chromatography. For example, for the mixed-mode separation method, Primesep B2 columns with embedded basic ion-pairing groups have been used for grapes, giving a significant improvement for chromatographic separation, purification, and detection of anthocyanidin diglucosides and acylated anthocyanins. However, the identification of anthocyanins is hard to predict by comparing with previously published data, because the ion-exchange elution mechanism significantly affects the retention orders of anthocyanins [56, 80].

The results from HPLC not only depend on column and separation mode but also on instruments and conditions used. Although instruments from different companies are comparable, the separation efficiency is always different among instruments. Once a new instrument is set up, it is better to optimize conditions and protocols for anthocyanins or other compounds. As such, results from the same instrument are highly reproducible. The detector can also provide informative characteristics of metabolites [82].

2.2 Detection, annotation, and identification

Ultraviolent (UV) and visible (Vis) detectors are commonly used in HPLC. The detectors measure the absorbance intensity of UV and Vis spectra between the 190 and 900 nanometer (nm) wavelengths. There are two types of UV-vis detectors, including tunable and photodiode array (PDA), also known as diode array (DAD) detectors. The tunable UV-vis detectors can measure the maximum absorption of each analyte of a sample at one or more discrete wavelengths during the analysis. PDA detectors can measure the absorbance of each analyte at the entire wavelength range or a fixed wavelength in real time (at intervals of 1 second or less) during separation by HPLC with continuous eluate delivery [83]. The use of a detector is dependent on the metabolites analyzed. To detect and/or identify anthocyanins, detectors mostly coupled with HPLC include UV-vis, /DAD, and PDA. These detector systems allow a few to full spectrophotometric scans on each peak as it elutes and provides a unique chromatogram for each anthocyanin that is used to compare with others for identification aims [6, 57]. In an acidic condition, the flavylium cation can maintain its red color and have absorption at a maximum between 510 and 545 nm (depending on the number of hydroxylated carbon atoms on the B-ring). The unique and maximum absorption wavelengths allow for distinguishing anthocyanins from other flavonoids for identification and quantification [6, 72]. PDA detector provides a spectral profile that can assist in detecting unknown peaks in the chromatograms, and provides characteristic spectra that give information about acylation and glycosylation patterns of anthocyanins [82, 84–86]. However, these detectors cannot distinguish anthocyanins with similar retention times and similar spectral characteristics. Also, the identification of anthocyanins with UV-vis dectors requires authentic standards, many of which are not commercially available [31, 82, 84, 87, 88].

Mass spectrometry (MS), an analytical technique, is used to measure the mass-to-charge ratio of ions. Mass spectrometry (MS) is completed on the mass spectrometer, which has MS detector. A mass spectrometer ionizes molecules and ionized molecules are sent to the mass analyzer, which is an electromagnetic field sorting and separating ions according to their mass and charge. Then, the mass detector detects and measures separated ions, and the results are displayed on a chart. HPLC coupled with a mass spectrometer has been used to effectively analyze anthocyanins (**Table 2**) and other plant secondary metabolites. Multiple accomplishments have been made to understand anthocyanins during fruit development [61, 89, 90]. In particular, HPLC-MS/MS or ultra-performance liquid chromatography (UPLC)-MS/MS is powerful to annotate unknown anthocyanins or identify known anthocyanins in fruits [61, 91]. These successes enhanced the understanding of anthocyanin biosynthesis and structures in fruits and other plant samples.

2.2.1 Electrospray ionization (ESI) mass spectrometry (MS): ESI-MS

The analysis of specific types of individual compounds by HPLC-MS requires an appropriate ionization interface between the physical coupling of LC and MS. Detection and quantification of an individual compound in MS-based measurements is determined by the level of ionization that generates intact molecular ions and/or a few fragments in MS1. Soft ionization techniques by desorption have been developed for nongaseous or thermally unstable natural compounds, for example, anthocyanins. These techniques cause a direct formation of gaseous ions by supplying power to solid or liquid sample, giving a little fragmentation and a simple mass spectrum for accurate molecular weight determination of the molecules. The most suitable ionization techniques for the chemical structure of anthocyanins are continuous-flow fast-atom bombardment (CF-FAB), desorption electrospray ionization (DESI), atmospheric pressure chemical ionization (APCI), matrix-assisted laser desorption ionization (MALDI), and electrospray ionization (ESI) [61, 82, 92–94].

In general, as an ionization mode, ESI has been mainly used, but some studies have reported the use of MALDI coupled with a time-of-flight (TOF) mass analyzer (MALDI-TOF) as an alternative. The main advantage of MALDI-TOF is the speed of analysis (a few minutes per sample). Also, MALDI-TOF mass spectrometry prevents the unwanted fragmentation of the molecules, giving a fingerprint mass spectrum for the desired molecules. Furthermore, this technique provides direct use of complex sample mixture without prior separation. For example, anthocyanin profiling of the crude aqueous-methanolic extract of the pulp of Jamun fruit was performed by MALDI-TOF mass spectrometry operating in positive ion mode and using sodium chloride and 2, 5-dihydroxybenzoic acid as the matrix. However, MALDI-TOF is not capable of generating MS/MS data compared to ESI-MS/MS systems [95, 96]. MALDI-TOF mass spectrometer was applied for analysis of anthocyanins from blueberries, and found to be quicker and to give nontargeted quantitative estimates compared to HPLC-PDA-MS method, but unable to distinguish between anthocyanins and other flavonoids, which generate ions of the same m/z value, giving an inherent limitation of the method [95, 97].

For MALDI mode, 2,4,6-trihydroxyacetophenon (THAP) and cyano-4-hydroxy-cinnamic acid (CHCA) are used as a potential matrix for the flavonoids, but these matrixes tend to be fragmented and decomposed under the most instrumental conditions, resulting in a complicated mass spectra and difficulties to analyze flavonoids with a small molecular mass. However, the surfactant cetyltrimethylammonium bromide (CTAB) has been introduced as a MALDI matrix-ion suppressor and reported to yield a higher resolution and greater reproducibility than those without surfactant for qualitatively identifying anthocyanins from multiple berry extracts in a few minutes. Because of the specificity of the matrix-ion suppression, the method is called "surfactant-mediated" MALDI, and demonstrated as a complementary rapid screening technique for anthocyanins [98].

ESI is an appropriate method to generate ions in a positive or negative mode. The only prerequisite that ESI ionization needs are that the sample of interest must be soluble in appropriate solvents, and introduced to a mass spectrometer in the form of a solution. In addition, ESI is a common interface between LC/MS because of avoiding many problems seen with other LC/MS ionization interfaces. ESI-MS is a powerful versatile ionization for thermally labile, nonvolatile, and polar compounds because this soft ionization technique can produce intact ions from large and complex compounds. Past studies have shown that ESI is effective for anthocyanin ionization that produces gaseous ions from highly charged evaporating liquid droplets. To date, ESI-MS has been described and used as a powerful technology to identify the molecular structure and contents of anthocyanins in fruits. Although both positive and negative ionization modes have been generally reported for analyses of fruit anthocyanins, the positive ionization mode has been more commonly preferred by researchers [61, 86, 99–101].

Atmospheric pressure chemical ionization (APCI) is another ionization interface that has been used particularly for the broad class of flavonoids such as aglycons and glycosides. Because APCI produces a single charged product, the molecular mass spectrum of the product can be directly observed. In contrast to ESI mode, LC coupled to positive or negative ion mode APCI is more suitable for the analysis of weakly polar or nonpolar compounds due to the sample vaporization [102, 103]. For identification of anthocyanins with the same molecular mass, either sample treatment such as acid hydrolysis must be performed to release the anthocyanidin aglycons or MS fragmentation data must be obtained. The mass spectrometric data of LC-APCI-MS method provides information on the fragmentation of the anthocyanins, allowing the identification of the conjugate and the aglycone moiety. Therefore, LC-APCI-MS method allows the characterization of anthocyanins in samples without the need for sample preparation. For example, anthocyanins from red raspberries have been identified from the methanolic extract by reversed-phase HPLC through an atmospheric pressure chemical ionization probe operating in positive ion mode [30]. The combination of both LC-APCI-MS and LC-ESI-MS methods was also reported to overcome the disadvantages of each ion source when applied individually, as well [103].

2.2.2 Tandem mass spectrometry (MS/MS)

Tandem mass spectrometry (MS/MS or MS^n) is powerful to characterize individual compounds and identify the structure of compounds by separate ionization and fragmentation steps. MS/MS allows for the formation of the fragments of each individual molecule by collision-induced dissociation (CID). Individual compounds are detected by the first quadrupole mass detector and then fragmented in the collision cell *via* a suitable gas, usually argon or nitrogen, and their fragments are detected by the second quadrupole mass analyzer [82]. In the last decade, the improvements in resolving power, selectivity, and sensitivity have accelerated the use of HPLC-MS/MS to identify known or annotate unknown anthocyanins [91].

2.3 HPLC coupled with ESI-quantitative time-of-flight MS/MS: HPLC/ESI-qTOF-MS/MS

For LC-MS interfaces, there are different types of mass analyzers available, such as magnetic sectors, time-of-flight (TOF) analyzers, quadrupole mass filters, quadrupole ion traps, and ion cyclotron resonance. Mass analyzers can broadly be divided into two main groups including high- and low-resolution analyzers depending on their ability to distinguish ions with small mass-to-charge (m/z) differences. The high-resolution analyzers are useful in the structural annotation of anthocyanins. They can provide accurate m/z values of fragments, which allows to predict the location of structural fissions in MS/MS fragmentations. The resulting fragments can be useful to annotate anthocyanin structures.

Time-of-flight mass analyzers work on the principle that lighter ions travel faster than heavier ions following an initial acceleration by an electric field. All ions acquire the same kinetic energy during this initial acceleration period and are separated in the field-free flight tube, according to their different velocities. The physical property that is measured is flight time, which is directly related to the mass-to-charge ratio of the ion. Due to this mode of operation, TOF instruments offer very high mass ranges, very high acquisition rates, relatively high resolving power, and good sensitivity [6, 76, 104].

Ion trap analyzers enable the true MS^n operation by allowing selective trapping and fragmentation of parent and/or daughter ions as a function of time. In general, ion trap analyzers are for qualitative analysis. The quadrupole or triple quadrupole ion traps are for quantitation purposes. For structure elucidation or annotation of an individual compound, a triple quadrupole ion trap has high specificity and sensitivity, thus it is an effective mass analyzer to filter the ion of choice. In the triple quadrupoles, the first quadrupole (Q1) and the third quadrupole (Q3) function as mass filters to isolate parent ions (precursor-ion) and to monitor selected daughter ions (fragment or product-ion), respectively. The second quadrupole (Q2) serves as a collision cell, where parent ions are fragmented by ionization [6, 76, 105–107]. For example, anthocyanins from northern highbush blueberry extracts were identified by performing purification of solid-phase extraction, elution with acidified water and methanol, separation with the gradient mix of acidified water and acetonitrile through Zorbax SB-C18 column, and detected by reversed-phase HPLC coupled with electrospray ionization probe operating in positive ion mode, and time-of-flight tandem mass spectrometer [55].

3. Conclusion

Anthocyanins are water-soluble glycosides acquiring different colors, from red to blue or violet. They are naturally the most occurring flavonoids containing sugar moiety and are synthesized abundantly in many fruits, particularly berries. Anthocyanins from fruits have high significance for food, cosmetic, and pharmaceutical industries. To obtain high-quality fruit anthocyanins, many different approaches, methods, and techniques have been created for extraction, structural characterization, and profiling. In this chapter, we introduced multiple methods for sample treatments, including extraction, chromatographic separation and purification, and for detection, annotation, and identification of fruit anthocyanins. We also discussed the use of HPLC-DAD in combination with mass spectrometry (MS) as an outstanding tool providing the chromatographic and spectral characteristics of the LC system and the resolution and separation by mass fragmentation. Especially, HPLC and MS/MS technologies were highlighted as powerful to understand anthocyanin profiles in fruits. As a result, the combination of the liquid chromatography (LC) method with electrospray ionization (ESI) and mass spectrometry (MS) or quadrupole time-of-flight (QTOF) with mass spectrometry (MS) was evaluated as the most popular and reliable methods for analyzing these compounds in fruit samples.

Acknowledgements

The authors acknowledge the Foreign Specialized Project Program of the Ministry of Science and Technology of the People's Republic of China (2021) for financial support through the granted project (Grant Number: QN2021016002L).

Conflict of interest

The authors declare no conflict of interest.

Author details

Seyit Yuzuak[1*], Qing Ma[2], Yin Lu[2] and De-Yu Xie[3]

1 Department of Molecular Biology and Genetic, Burdur Mehmet Akif Ersoy University, Burdur, Turkey

2 College of Biology and Environmental Engineering, Zhejiang Shuren University, Zhejiang, Hangzhou, China

3 Department of Plant and Microbial Biology, North Carolina State University, Raleigh, NC, USA

*Address all correspondence to: syuzuak@mehmetakif.edu.tr

References

[1] Fossen T et al. Anthocyanins of grasses. Biochemical Systematics and Ecology. 2002;**30**(9):855-864

[2] Fabroni S et al. Screening of the anthocyanin profile and in vitro pancreatic lipase inhibition by anthocyanin-containing extracts of fruits, vegetables, legumes and cereals. Journal of the Science of Food and Agriculture. 2016;**96**(14):4713-4723

[3] Passeri V, Koes R, Quattrocchio FM. New challenges for the Design of High Value Plant Products: Stabilization of anthocyanins in plant vacuoles. Front. Plant Science. 2016;**7**:153

[4] Kong JM et al. Analysis and biological activities of anthocyanins. Phytochemistry. 2003;**64**(5):923-933

[5] Harborne JB, Williams CA. Advances in flavonoid research since 1992. Phytochemistry. 2000;**55**(6):481-504

[6] Welch CR, Wu QL, Simon JE. Recent advances in anthocyanin analysis and characterization. Current Analytical Chemistry. 2008;**4**(2):75-101

[7] Delgado-Vargas F, Jimenez AR, Paredes-Lopez O. Natural pigments: Carotenoids, anthocyanins, and betalains - characteristics, biosynthesis, processing, and stability. Critical Reviews in Food Science and Nutrition. 2000;**40**(3):173-289

[8] He J, Giusti MM. Anthocyanins: Natural colorants with health-promoting properties. Annual Review of Food Science and Technology. 2010;**1**:163-187

[9] Wrolstad RE. Symposium 12: Interaction of natural colors with other ingredients - anthocyanin pigments - bioactivity and coloring properties. Journal of Food Science. 2004;**69**(5):C419-C421

[10] Winefield C. The final steps in anthocyanin formation: A story of modification and sequestration. Advances in Botanical Research;**372002**:55-74

[11] Furuta K. Host preference and population-dynamics in an autumnal population of the maple aphid, Periphyllus-Californiensis Shinji (Homoptera, Aphididae). Journal of Applied Entomology-Zeitschrift Fur Angewandte Entomologie. 1986;**102**(1):93-100

[12] Page JE, Towers GHN. Anthocyanins protect light-sensitive thiarubrine phototoxins. Planta. 2002;**215**(3):478-484

[13] Ferreyra MLF, Rius SP, Casati P. Flavonoids: biosynthesis, biological functions, and biotechnological applications. Front. Plant Science. 2012;**3**:222

[14] Sun Y, Li H, Huang JR. Arabidopsis TT19 functions as a carrier to transport anthocyanin from the cytosol to tonoplasts. Molecular Plant. 2012;**5**(2):387-400

[15] Winkel-Shirley B, Flavonoid biosynthesis. A colorful model for genetics, biochemistry, cell biology, and biotechnology. Plant Physiology. 2001;**126**(2):485-493

[16] Xie DY, Dixon RA. Proanthocyanidin biosynthesis - still more questions than answers? Phytochemistry. 2005;**66**(18):2127-2144

[17] Poracova J et al. The importance of anthocyanins for human and animal health. Planta Medica. 2011;**77**(12):1447-1447

[18] Galvano F et al. Cyanidins: Metabolism and biological properties. Journal of Nutritional Biochemistry. 2004;**15**(1):2-11

[19] Matsumoto H et al. Stimulatory effect of cyanidin 3-glycosides on the regeneration of rhodopsin. Journal of Agricultural and Food Chemistry. 2003;**51**(12):3560-3563

[20] Folts JD. Antithrombotic potential of grape juice and red wine for preventing heart attacks. Pharmaceutical Biology. 1998;**36**:21-27

[21] Keevil JG et al. Grape juice inhibits human *ex vivo* platelet aggregation while orange and grapefruit juices do not. Journal of the American College of Cardiology. 1998;**31**(2):172a-172a

[22] Tsuda T et al. Dietary cyanidin 3-O-beta-D-glucoside-rich purple corn color prevents obesity and ameliorates hyperglycemia in mice. Journal of Nutrition. 2003;**133**(7):2125-2130

[23] Xie LH et al. Recent advances in understanding the anti-obesity activity of anthocyanins and their biosynthesis in microorganisms. Trends in Food Science & Technology. 2018;**72**:13-24

[24] Hou DX et al. Molecular mechanisms behind the chemopreventive effects of anthocyanidins. Journal of Biomedicine and Biotechnology. 2004;**5**:321-325

[25] Francisco RM et al. ABCC1, an ATP binding cassette protein from grape berry, transports Anthocyanidin 3-O-glucosides. Plant Cell. 2013;**25**(5):1840-1854

[26] Martín J et al. Anthocyanin Pigments: Importance, Sample Preparation and Extraction. Phenolic Compounds - Natural Sources, Importance and Applications, InTech, Crossref. Mar 2017. DOI: 10.5772/66892

[27] Yuzuak S, Xie DY. Anthocyanins from muscadine (Vitis rotundifolia) grape fruit. Current. Plant Biology. 2022;**30**:100243

[28] Prior RL et al. Identification of procyanidins and anthocyanins in blueberries and cranberries (Vaccinium spp.) using high-performance liquid chromatography/mass spectrometry. Journal of Agricultural and Food Chemistry. 2001;**49**(3):1270-1276

[29] Stintzing FC et al. A novel zwitterionic anthocyanin from evergreen blackberry (Rubus laciniatus Willd.). Journal of Agricultural and Food Chemistry. 2002;**50**(2):396-399

[30] Mullen W, Lean MEJ, Crozier A. Rapid characterization of anthocyanins in red raspberry fruit by high-performance liquid chromatography coupled to single quadrupole mass spectrometry. Journal of Chromatography. A. 2002;**966**(1-2):63-70

[31] Kim M-Y et al. Identification and antiradical properties of anthocyanins in fruits of Viburnum dilatatum thunb. Journal of Agricultural and Food Chemistry. 2003;**51**(21):6173-6177

[32] Määttä K, Kamal-Eldin A, Törrönen AR. High-performance liquid chromatography (HPLC) analysis of phenolic compounds in berries with diode array and electrospray ionization mass spectrometric (MS) detection: Ribes species. Journal of Agricultural and Food Chemistry. 2003;**51**(23):6736-6744

[33] Nakajima J-I et al. LC/PDA/ESI-MS profiling and radical scavenging activity of anthocyanins in various berries. Journal of Biomedicine and Biotechnology. 2004;**2004**:241-247

[34] Kammerer D et al. Polyphenol screening of pomace from red and white

grape varieties (Vitis vinifera L.) by HPLC-DAD-MS/MS. Journal of Agricultural and Food Chemistry. 2004;**52**(14):4360-4367

[35] Määttä-Riihinen KR, Kamal-Eldin A, Törrönen AR. Identification and quantification of phenolic compounds in berries of Fragaria and Rubus species (family Rosaceae). Journal of Agricultural and Food Chemistry. 2004;**52**(20):6178-6187

[36] Longo L, Vasapollo G. Determination of anthocyanins in Ruscus aculeatus L. berries. Journal of Agricultural and Food Chemistry. 2005;**53**(2):475-479

[37] Longo L, Vasapollo G, Rescio L. Identification of anthocyanins in Rhamnus alaternus L. berries. Journal of Agricultural and Food Chemistry. 2005;**53**(5):1723-1727

[38] Montefiori M et al. Pigments in the fruit of red-fleshed kiwifruit (Actinidia chinensis and Actinidia deliciosa). Journal of Agricultural and Food Chemistry. 2005;**53**(24):9526-9530

[39] Zanatta CF et al. Determination of anthocyanins from camu-camu (*Myrciaria dubia*) by HPLC-PDA, HPLC-MS, and NMR. Journal of Agricultural and Food Chemistry. 2005;**53**(24):9531-9535

[40] Wu X, Prior RL. Systematic identification and characterization of anthocyanins by HPLC-ESI-MS/MS in common foods in the United States: Fruits and berries. Journal of Agricultural and Food Chemistry. 2005;**53**(7):2589-2599

[41] Kahle K et al. Studies on apple and blueberry fruit constituents: Do the polyphenols reach the colon after ingestion? Molecular Nutrition & Food Research. 2006;**50**(4-5):418-423

[42] Nicoué EE, Savard S, Belkacemi K. Anthocyanins in wild blueberries of Quebec: Extraction and identification. Journal of Agricultural and Food Chemistry. 2007;**55**(14):5626-5635

[43] Jin H et al. Characterization of anthocyanins in wild Lycium ruthenicum Murray by HPLC-DAD/QTOF-MS/MS. Analytical Methods. 2015;**7**:4947-4956

[44] Sharma RJ et al. Metabolite fingerprinting of Eugenia jambolana fruit pulp extracts using NMR, HPLC-PDA-MS, GC-MS, MALDI-TOF-MS and ESI-MS/MS spectrometry. Natural Product Communications. 2015:**10**(6):969-976

[45] Ramirez JE et al. Anthocyanins and antioxidant capacities of six Chilean berries by HPLC-HR-ESI-ToF-MS. Food Chemistry. 2015;**176**:106-114

[46] Legua P et al. Polyphenolic compounds, anthocyanins and antioxidant activity of nineteen pomegranate fruits: A rich source of bioactive compounds. Journal of Functional Foods. 2016;**23**:628-636

[47] Oszmiański J, Lachowicz S. Effect of the production of dried fruits and juice from chokeberry (Aronia melanocarpa L.) on the content and Antioxidative activity of bioactive compounds. Molecules. 2016;**21**(8):1098

[48] Scorrano S et al. Anthocyanins profile by Q-TOF LC/MS in Myrtus communis berries from Salento area. Food Analytical Methods. 2017;**10**:2404-2411

[49] Li F et al. The novel contributors of anti-diabetic potential in mulberry polyphenols revealed by UHPLC-HR-ESI-TOF-MS/MS. Food Research International. 2017;**100**(Pt 1):873-884

[50] Sang J et al. Extraction and characterization of anthocyanins from

Nitraria tangutorun bobr. Dry fruit and evaluation of their stability in aqueous solution and taurine-contained beverage. Journal of Food Measurement and Characterization. 2017;**12**:937-948

[51] Pastrana-Bonilla E, Akoh CC, Cerquera NE. Identification and quantification of anthocyanins in muscadine grapes by HPLC and HPLC-MS. ARPN Journal of Engineering and Applied Sciences. 2017;**12**(2):626-631

[52] Wu Q et al. Analysis of polyphenols composition and antioxidant activity assessment of Chinese dwarf cherry (Cerasus humilis (Bge.) Sok.). Natural Product Communications. 2019;**14**(6)

[53] Li CX et al. Phytochemical profiles, antioxidant, and antiproliferative activities of red-fleshed apple as affected by in vitro digestion. Journal of Food Science. 2020;**85**(9):2952-2959

[54] Stanoev JP, Stefova M, Bogdanov JB. Systematic HPLC/DAD/MSn study on the extraction efficiency of polyphenols from black goji: Citric and ascorbic acid as alternative acid components in the extraction mixture. Journal of Berry Research. 2021;**11**(269):1-20

[55] Pico J et al. Determination of free and bound phenolics in northern highbush blueberries by a validated HPLC/QTOF methodology. Journal of Food Composition and Analysis. 2022;**108**:104412

[56] Liao Z et al. Recovery of value-added anthocyanins from mulberry by a cation exchange chromatography. Current Research in Food Science. 2022;**5**:1445-1451

[57] Mazza G, Cacace J, Kay C. Methods of analysis for anthocyanins in plants and biological fluids. Journal of AOAC International. 2004;**87**:129-145

[58] Horbowicz M et al. Anthocyanins of fruits and vegetables-their occurrence, analysis and role in human nutrition. Vegetable Crops Research Bulletin. 2008;**68**:5-22

[59] Avula B et al. Advances in the chemistry, analysis and adulteration of anthocyanin rich-berries and fruits: 2000-2022. Molecules. 2023;**28**(2):560. DOI: 10.3390/molecules28020560

[60] Lopez E et al. Effect of light on stability of anthocyanins in Ethanolic extracts of Rubus fruticosus. Food and Nutrition Sciences. 2014;**05**:488-494

[61] da Silva Oliveira JP et al. Metabolomic studies of anthocyanins in fruits by means of a liquid chromatography coupled to mass spectrometry workflow. Current. Plant Biology. 2022;**32**:100260

[62] Tan J et al. Extraction and purification of anthocyanins: A review. Journal of Agriculture and Food Research. 2022;**8**:100306

[63] Nunes AN et al. Alternative extraction and downstream purification processes for anthocyanins. Molecules. 2022;**27**(2):368

[64] Gonzalez-Paramas AM et al. Analysis of flavonoids in foods and biological samples. Mini-Reviews in Medicinal Chemistry. 2011;**11**(14):1239-1255

[65] de Rijke E et al. Analytical separation and detection methods for flavonoids. Journal of Chromatography A. 2006;**1112**(1-2):31-63

[66] Flamini R. Recent applications of mass spectrometry in the study of grape and wine polyphenols. International Scholarly Research Notices. 2013;**2013**:1-45

[67] Gardana C et al. Bilberry adulteration: Identification and

chemical profiling of anthocyanins by different analytical methods. Journal of Agricultural and Food Chemistry. 2014;**62**(45):10998-11004

[68] Fiorelia NE et al. Types of high-performance liquid chromatography (HPLC) columns: A review. FoodTech: Jurnal Teknologi Pangan. 2022;**5**(1):1-16

[69] Chen Y et al. Adsorption properties of macroporous adsorbent resins for separation of anthocyanins from mulberry. Food Chemistry. 2016;**194**:712-722

[70] Berthod A. Countercurrent chromatography: The support-free liquid stationary phase. Elsevier Science; 2002;**38**:1-397

[71] Petersson EV et al. Analysis of anthocyanins in red onion using capillary electrophoresis-time of flight-mass spectrometry. Electrophoresis. 2008;**29**(12):2723-2730

[72] da Costa CT, Horton D, Margolis SA. Analysis of anthocyanins in foods by liquid chromatography, liquid chromatography-mass spectrometry and capillary electrophoresis. Journal of Chromatography A. 2000;**881**(1-2):403-410

[73] Calvo D et al. Migration order of wine anthocyanins in capillary zone electrophoresis. Analytica Chimica Acta. 2004;**524**(1-2):207-213

[74] Saenz-Lopez R, Fernandez-Zurbano P, Tena MT. Development and validation of a capillary zone electrophoresis method for the quantitative determination of anthocyanins in wine. Journal of Chromatography A. 2003;**990**(1-2):247-258

[75] Bednář P et al. Utilization of capillary electrophoresis/mass spectrometry (CE/MSn) for the study of anthocyanin dyes. Journal of Separation Science. 2005;**28**(12):1291-1299

[76] de Villiers A, Venter P, Pasch H. Recent advances and trends in the liquid-chromatography-mass spectrometry analysis of flavonoids. Journal of Chromatography A. 2016;**1430**:16-78

[77] Buszewski B, Noga S. Hydrophilic interaction liquid chromatography (HILIC)—A powerful separation technique. Analytical and Bioanalytical Chemistry. 2011;**402**:231-247

[78] Willemse CM, Stander MA, de Villiers A. Hydrophilic interaction chromatographic analysis of anthocyanins. Journal of Chromatography. A. 2013;**1319**:127-140

[79] Markham KA, Kohen A. Analytical procedures for the preparation, isolation, analysis and preservation of reduced nicotinamides. Current Analytical Chemistry. 2006;**2**(4):379-388

[80] McCallum J et al. Improved high performance liquid chromatographic separation of anthocyanin compounds from grapes using a novel mixed-mode ion-exchange reversed-phase column. Journal of Chromatography. A. 2007;**1148**:38-45

[81] Pismenskaya N et al. Adsorption of anthocyanins by cation and anion exchange resins with aromatic and aliphatic polymer matrices. International Journal of Molecular Sciences. 2020;**21**:1-26

[82] Giusti MM et al. Electrospray and tandem mass spectroscopy as tools for anthocyanin characterization. Journal of Agricultural and Food Chemistry. 1999;**47**(11):4657-4664

[83] Swartz ME. HPLC detectors: A brief review. Journal of Liquid

Chromatography & Related Technologies. 2010;**33**:1130-1150

[84] Hong V, Wrolstad RE. Use of Hplc separation photodiode Array detection for characterization of anthocyanins. Journal of Agricultural and Food Chemistry. 1990;**38**(3):708-715

[85] Bakker J, Timberlake CF. The distribution of anthocyanins in grape skin extracts of port wine cultivars as determined by high-performance liquid-chromatography. Journal of the Science of Food and Agriculture. 1985;**36**(12):1315-1324

[86] Garcia-Beneytez E, Cabello F, Revilla E. Analysis of grape and wine anthocyanins by HPLC-MS. Journal of Agricultural and Food Chemistry. 2003;**51**(19):5622-5629

[87] Gao L, Mazza G. Rapid method for complete chemical characterization of simple and Acylated anthocyanins by high-performance liquid-chromatography and capillary gas-liquid-chromatography. Journal of Agricultural and Food Chemistry. 1994;**42**(1):118-125

[88] Strack D, Akavia N, Reznik H. High-performance liquid-chromatographic identification of anthocyanins. Zeitschrift Fur Naturforschung C-a Journal of Biosciences. 1980;**35**(7-8):533-538

[89] Baldi A et al. Hplc/Ms application to anthocyanins of Vitis-Vinifera L. Journal of Agricultural and Food Chemistry. 1995;**43**(8):2104-2109

[90] Wang J, Sporns P. Analysis of anthocyanins in red wine and fruit juice using MALDI-MS. Journal of Agricultural and Food Chemistry. 1999;**47**(5):2009-2015

[91] Saha S et al. Anthocyanin profiling using UV-Vis spectroscopy and liquid chromatography mass spectrometry. Journal of AOAC International. 2019;**103**(1):23-39

[92] Sagesser M, Deinzer M. HPLC-ion spray tandem mass spectrometry of flavonol glycosides in hops. Journal of the American Society of Brewing Chemists. 1996;**54**(3):129-134

[93] Aramendia MA et al. Determination of Isoflavones using capillary electrophoresis in combination with electrospray mass-spectrometry. Journal of Chromatography A. 1995;**707**(2):327-333

[94] Favretto D, Flamini R. Application of electrospray ionization mass spectrometry to the study of grape anthocyanins. American Journal of Enology and Viticulture. 2000;**51**(1):55-64

[95] Wang J, Kalt W, Sporns P. Comparison between HPLC and MALDI-TOF MS analysis of anthocyanins in highbush blueberries. Journal of Agricultural and Food Chemistry. 2000;**48**:3330-3335

[96] Castañeda A et al. Identification of anthocyanins in red grape, plum and capulin by MALDI-ToF MS. Journal of the Mexican Chemical Society. 2012;**2012**:378-383

[97] Mullen W et al. Use of accurate mass full scan mass spectrometry for the analysis of anthocyanins in berries and berry-fed tissues. Journal of Agricultural and Food Chemistry. 2010;**58**(7):3910-3915

[98] Grant D, Helleur R. Rapid screening of anthocyanins in berry samples by surfactant-mediated matrix-assisted laser desorption/ionization time-of-flight mass spectrometry. Rapid communications in mass spectrometry: RCM. 2008;**22**:156-164

[99] Fenn JB et al. Electrospray ionization for mass-spectrometry

of large biomolecules. Science. 1989;**246**(4926):64-71

[100] Hutton T, Major HJ. Characterizing biomolecules by electrospray ionization mass spectrometry coupled to liquid chromatography and capillary electrophoresis. Biochemical Society Transactions. 1995;**23**(4):924-927

[101] Dugo P et al. Identification of anthocyanins in berries by narrow-bore high-performance liquid chromatography with electrospray ionization detection. Journal of Agricultural and Food Chemistry. 2001;**49**(8):3987-3992

[102] de Rijke E et al. Liquid chromatography with atmospheric pressure chemical ionization and electrospray ionization mass spectrometry of flavonoids with triple-quadrupole and ion-trap instruments. Journal of Chromatography A. 2003;**984**(1):45-58

[103] Commisso M et al. Performance comparison of electrospray ionization and atmospheric pressure chemical ionization in untargeted and targeted liquid chromatography/mass spectrometry based metabolomics analysis of grapeberry metabolites. Rapid communications in mass spectrometry: RCM. 2017;**31**(3):292-300

[104] Ignat I, Volf I, Popa VI. A critical review of methods for characterisation of polyphenolic compounds in fruits and vegetables. Food Chemistry. 2011;**126**(4):1821-1835

[105] Flamini R. Mass spectrometry in grape and wine chemistry. Part I: Polyphenols. Mass Spectrometry Reviews. 2003;**22**(4):218-250

[106] Huang ZL et al. Identification of anthocyanins in muscadine grapes with HPLC-ESI-MS. Lwt-Food Science and Technology. 2009;**42**(4):819-824

[107] Ganzera M, Sturm S. Recent advances on HPLC/MS in medicinal plant analysis-an update covering 2011-2016. Journal of Pharmaceutical and Biomedical Analysis. 2018;**147**:211-233

Chapter 8

Current Trends in HPLC for Quality Control of Spices

Prafulla Kumar Sahu, Sukumar Purohit, Swarnajeet Tripathy, Durga Prasad Mishra and Biswajeet Acharya

Abstract

India, the land of spices and condiments, is endowed with a plethora of herbs, spices, and unusual plants. Spices have been used as flavoring and coloring agents in Indian society since time immemorial. Spices have also been shown to have antioxidant, antibacterial, anticancer, and anti-inflammatory properties. Assessing spices' taste, nutritional, and bioactive qualities during postharvest processing is critical for quality control and preventing adulteration. Various illegal colors are frequently used to adulterate spices for fraudulent trading operations. For instance, Sudan dyes are widely substituted with hot chili, red pepper, or tomato products; metanil yellow in turmeric; tartrazine, amaranth, and sunset yellow FCF in ginger and chili powder; and magenta III and rhodamine B in saffron. These adulterants degrade the flavoring, fragrance, cosmetics, medicinal, and preservative value of spices, their authentication is critical in quality control. Apart from these adulterants, various aflatoxins secreted after fungal contamination also cause quality degradation of spices. According to the literature evaluation, HPLC is a rapid and adaptable technique for efficiently identifying these compounds in spices. The proposed chapter summarizes application of HPLC for detection, quantification, and quality assessment of various spices. Some of the recently published work on the said topic from various search engines (Google scholar, Scopus, science direct, etc.) is mentioned in the chapter.

Keywords: analytical HPLC, spice and condiments, adulteration, aflatoxin, quality control

1. Introduction

Spice consumption has been a long-standing habit due to the great value of its color, flavor, pungency, and aroma properties. Spices are rich in lipids, proteins, minerals, and vitamins, in addition to their organoleptic qualities [1]. In addition, they are effective against microorganisms, oxidative stress, inflammation, diabetes, immunosuppressant, and mutation [2–5]. They are excellent for food preservation. As a result of its many advantageous effects, including the ability to purify blood and condition the skin, these spices have been mentioned in the ancient system of medicine such as Ayurveda, Unani, and Homeopathy. Besides the health benefits, global food habits include utilization of high spice levels, which makes the food more palatable and create an eye-catching garnish.

IntechOpen

Global herb and spice market is valued at four billion USD and is believed to further grow up to 6–6.5 billion USD in next decade [6]. The increasing demand for spices is because of their flavor, aroma, taste, and color. Many of these spices and herbs include turmeric, ginger, garlic, chili, pepper, etc. used in every household worldwide. Apart from these whole spices, their powders (chili powder, turmeric powder, ginger powder, pepper powder, etc.) are also used for seasoning. Unfortunately, these powders are often contaminated with chalk powder, dyes, and many other chemicals to enhance the bulk and colored texture of the spice [7]. This food fraud practice can be detrimental to health condition of consumers. For instance, various reports suggest that nut protein mixed with cumin can cause anaphylaxis [8]. Yellow chalk powder is often mixed with turmeric powder, which may cause severe nausea, vomiting, and loss of appetite to the consumers [9]. Mixing olive leaves with oregano is another example of indirect type of food fraud, which can cost toxicity and mutagenicity to the customer [10].

Intentional addition or substitution of a substance with a structurally similar substance to enhance its quantity and decrease its production and processing cost for economic gain is called economically motivated adulteration [11]. This practice encourages fraudsters to mix various harmful illegal dyes with spices to make them more appealing to the customer. For instance, Sudan one and Sudan four are mixed with turmeric, chili, curry, pepper, etc. [12]. Rhodamine B is mixed with paprika, sumac, chili, etc. [13]. Para red is an illegal dye often mixed with cayenne pepper, chili, and paprika to enhance the color of the spice [13]. These above-mentioned banned dyes are genotoxic and carcinogenic in nature. Therefore, identifying these adulterants from herbs and spices has become an important step for their quality control.

Owing to the huge economic potential of the spice market, herbs and spices are heavily adulterated. Very often they are adulterated with low-quality and substandard products. This practice further enhances the possibilities of contamination. Improper storage and handling have a significant role in quality degradation of spice, which includes microbial growth [14]. Besides bacterial contamination, spices are more susceptible to fungal contamination. Mycotoxins are toxic materials released to the fungus-contaminated spices and herbs. Aflatoxins are a type of mycotoxins that are produced by certain fungi causing severe contamination in many agricultural crops including spices [15]. Aflatoxin contamination in spices occurs during their harvesting, drying, and storage [16]. These toxins must be identified and approaches should be considered to minimize such contamination for the quality enhancement of spices.

To verify the authenticity of herbs and spices due to the surging trends in spice contamination, many analytical techniques are taken into consideration. Most relevant techniques include spectroscopy and chromatography [17]. Visual inspection and microscopic methods are also considered, but they require huge manpower, trained professionals, and more analysis time. Many DNA-based methods, such as random amplified polymorphic DNA (RAPD), are also used for the detection of adulteration in spices. However, the authenticity of the result reproducibility is a matter of discussion [18]. High-performance liquid chromatography (HPLC) is a rapid, reliable, accurate, sensitive, and highly specific technique considered the gold standard for the detection of adulterants [19]. This chapter gives a complete overview of the ancient use of spices in India and their trading and fraudulent activities followed over the years to adulterate spices. Further, the report summarizes various recent findings based on identification of different adulterants including synthetic dyes, herbicides and pesticides, starch and fillers, etc. through HPLC methods that are mentioned. In addition to this, role of HPLC in quality assessment of drugs bearing essential oil and mycotoxins has been discussed elaborately. A comparison study between HPLC analysis and other spectroscopic and chromatographic techniques is also discussed in this manuscript. This chapter will be the first documentation of its kind to cover the role of HPLC in quality boosting of different herbs and spices.

1.1 History, trivia, and background about spices of India

Spices have various purposes, including cooking, aroma, personal care, medicine, and long-term food preservation. Most countries that import goods find spices important. There are many aromatic spices and flavors in Indian history. Spices have been pivotal in many realms, from mythology and the Middle Ages to modern politics and economics. Spices were used as exchange and gifts for marriages, war treaties, political tradings, etc. [20–22].

Spices have long been a globally traded commodity. Spices were one of the most important components of trade from the Indian subcontinent to the Roman Empire throughout the first to third centuries CE [23]. The Middle East initially used spices circa 5000 BC, which further moved through Egyptians exchanged between 3000 to 200 BC. They were cultivating garlic to use as a fragrance in the process of mummification. The Romans dominated the trade of spices including nutmeg, cinnamon, pepper, cloves, and ginger from 200 BC until 1200 AD. According to Indian history, in 1497, Portuguese explorer Vasco de Gama lured princes of India with spices such as ginger, pepper, and cinnamon. In 1663 AD, the Dutch people acquired exclusive permission for pepper trading with India and controlled Asian spices by the 17th century. In the late 17th century, the French became a superpower and stole cloves, cinnamon, and nutmeg from the Dutch. In 1780–1799, British took seized spice trading centers. By the year 1672 America joined the spice race through its geopolitical and economic eminence.

India is the world's greatest spice grower, accounting for 75% of worldwide spice production. The overseas trading of spices by India in 2018–2019 was 0.85 million tonnes ($2.25 billion). The spice market's growth in recent years is a testament to the popularity of Indian spices abroad. Flavor, color, aroma, preservation, and therapeutic characteristics give spices considerable economic worth. Important seasonings, tree spices, seed spices, and miscellaneous spices all make up the spice family [24]. It is estimated that between 6.8 and 12.5 percent of spices are lost after harvest [25]. Spice businesses also experience pollution and adulteration. Adulteration causes health dangers, dangerous products, and poor quality.

The spice sector is constantly challenged by adulteration, which can take the form of introducing low-grade, harmful, or low-quality commodities as well as extraneous chemicals. Adulteration not only reduces the overall quality and authenticity of a product but it can also have a negative impact on consumer health, including the development of long-term conditions such as paralysis, cancer, and a compromised immune system because of the presence of poisonous products inside the human body [26]. Spice adulteration business stands at around $30–$40 billion per year as per the estimate given by Global Food Safety Forum. Adulteration is done to profit from cheaper raw materials and meets overpopulation's food need. Lack of knowledge about adulterants in spices lowers their marketability and safety.

1.2 Importance of spices for the Indian economy

Spices are popular everywhere. India is the largest exporter and producer of herbs and spices worldwide. Spices are a highly traded commodity. Spice imports from developing countries like India dominate the worldwide market [27].

In India, the best quality spices are cultivated in the western ghat, Coorg, and Malabar region of southern states. Apart from that, other Indian states such as Madhya Pradesh, Kashmir, and Uttar Pradesh depend on their geo-climatic circumstances. In Uttar Pradesh, cumin, coriander, fennel, and black seed are commercially grown [28]. Spices are an ideal crop for small-scale farming in India. This firm may provide the family with more career options and emergency cash. Spices are a good business for women since they may be cultivated in home gardens, and they contribute to the local economy.

India's spices are the world's best. Black pepper has driven trade policy since time immemorial [29]. Spice trade fortunes affect agriculture exports. Spices accounted for 8.4% of all agricultural exports in 2017–2018, and their total worth was $4.69 billion in 2016 [30].

1.3 Scope of adulteration in the spice quality

Adulteration uses forbidden substances such as sand, pebbles, woodchip, colors, oil, floral stalks, and so on to improve the physical appearance of the spices. Lead and arsenic contamination can be developed during food preparation and handling [26]. Ground spices are more likely adulterated because of their shape and texture, which can be easily admixed during grinding or milling. Spice adulteration is done during processing of the spices or transpiration for the benefit of the trader. Common adulteration practices for different spices are presented in **Figure 1**.

Authenticity of food products, herbs, and spices has become a significant step of their quality control. Today, the consumers rely more on 100% safer and natural products. Therefore, identification of adulterants mixed with spices has become very important for checking the authenticity of the spices. Pure and authentic product assures the high quality of the product to the purchaser, dealer, and exporter. To authenticate the originality on the scale of quality and standard index, the analytical

Figure 1.
Example of some household spices with their adulterants. Red brick dust (a) is mixed with red chili powder (b), metanil yellow (c) is mixed with turmeric powder (d), papaya seeds (e) is mixed with black pepper (f), grass seed (g) is used as adulterant for cumin seed (h), cassia (i) is used as a substitution for cinnamon (j), corn threads (k) are mixed with saffron (l), de-oiled cardamom (m) is mixed with good quality cardamom for bulk (n), agremon seed (o) is often mixed with mustard seed (p), and soap stone (q) is mixed with asafoetida (r).

approaches are classified into three major basic strategies: physical, chemical, and instrumental analysis. Though physical methods are simple but claim limited applications due to time consumption. Chemical and instrumental techniques have been widely used. Although these techniques involve complex instrumentation and data processing, their routine applications cannot be restricted owing to their powerful benefits of rapidity, sensitivity, accuracy, and cost-effectiveness.

1.4 Role of HPLC in detecting adulterants in spices

Chromatography-based methods successfully separate a mixture of components. It has been used for identifying secondary plant metabolites for ages [31, 32]. However, it is a popular approach for detecting and identifying food adulterants also. Based on its principle, HPLC separates chemical entities from a sample mixture based on their affinity for the column adsorbent or mobile phase, causing constituents to flow at various speeds and separate. It was once called as high-pressure liquid chromatography because it used high-pressure pumps [33]. There are many toxic materials and banned dyes; adulterant compositions have been identified through HPLC analysis. For instance, Bhooma et al. [34] reported presence of magenta III and rhodamine B in pink saffron. They collected 104 commercial saffron samples from 16 different countries and confirmed presence of magenta III and rhodamine B in 20 samples for the first time. Both the toxic dyes were identified by HPLC and ESI MS analysis. Further, Sahu et al. [35] used reversed-phase HPLC for simultaneous determination and separation of curcumin, metanil yellow, demethoxy curcumin, and bismethoxy curcumin. Metanil yellow is a carcinogenic and genotoxic banned dye often used as an adulterant for turmeric powder. The authors reported that the RP-HPLC was very accurate and precise to detect turmeric adulterants with a detection limit of 0.37–2.48 mg/ml concentration. Another report by Vickers et al. suggested that a minimum limit of quantification (0.1 mg/kg) was achieved using HPLC for the detection of genotoxic Sudan dyes (I, II, III, and IV) through HPLC analysis from different spices from Egypt [36]. Adulteration in spices alone cannot deteriorate their quality aspects. Improper harvesting, processing, drying, and storage lead to poor quality of spices [37]. Inadequate measures followed for drying spices promote higher moisture content and further encourage microbial contamination. Spices are prone to get contaminated with toxic fungi and molds. Fungal species such as *Aspergillus* and *Penicillium* release toxic secondary metabolites such as mycotoxin. Aflatoxins are a group of mycotoxins that are potentially dangerous when released into spices and further degrade the food material. However, recent chromatographic techniques such as HPLC have become a go-to tool for detection of such toxic material in spices. Mixing of starch and other harmful powder in powdered drugs has also become a concern these days. However, some recent reports suggest HPLC plays a vital role in detecting such type of misconduct. This chapter summarizes some of the recent literature where HPLC has been employed for the detection of adulterants such as dyes, mycotoxins, pesticides, and powder fillers in spices.

2. Identification of synthetic dyes using HPLC

Synthetic dyes are commonly used to enhance the color of spices such as chili powder, paprika, and turmeric. While some of these dyes are approved for use in food, others may be unsafe and pose a risk to human health. High-performance liquid

Spices	Adulterant	Extraction process	Detection	LC condition	Selection criteria	Ref.
Chili Pepper	Sudan I, II, III, and IV	Vortex and centrifugation with ACN	PDA	Isocratic; Varian Microsorb-MV RP C18 column (150 × 4.6 mm, 100^{-5}); mobile phase ACN: water (80:20, v/v); flow rate 1 mL/min	Simple and fast method LOD: 1.5–2 mg/kg LOQ: 3–4 mg/kg	[44]
Chili Pepper	Sudan I, II, III, and IV	SPE	DAD & MS/MS	Isocratic; Agilent Eclipse 5 μm XDB-C18 (4.6 × 150 mm); mobile phase methanol: water (95: 5 v/v) buffered with 5 mM ammonium formate and 0.1% formic acid; flow rate 0.8 ml/ min	Reproducible and selective LOD: 1.5–2 mg/kg	[45]
Sumac, hot chili, paprika, cumin, and turmeric	Sudan I, II, III, IV, Para Red, Orange G, and Red 7B	Shaken with ACN → centrifugation	DAD	Isocratic; Agilent Eclipse® plus C18 (4.6 × 250 mm 5 μm); mobile phase ACN: water (95: 5 v/v); flow rate 1 ml/min	Sensitive and selective LOD: 0.01 to 0.03 mg/kg LOQ: 0.1 mg/kg	[46]
Varieties of anise, black pepper, cardamom, cayenne pepper, chili, cinnamon, clove, coriander, cumin, ginger, liquorice, mace, mustard	58 illegal dyes	Water: ACN (60:40) → mechanical shake → centrifugation	MS/MS	Linear gradient; Waters Acquity BEH Shield RP18 column (2.1 × 100 mm, 1.7 μm); mobile phase 20 mM ammonium formate and ACN; flow rate 0.4 mL/min	Accurate LOD: 0.005–2.0 mg/kg	[47]
Branded and non-branded multiple spices (biryani, Chaat and 16 other spices)	Sudan I, II, III, and IV	Soaked in ethyl acetate overnight at 4°C → extracted color by rotary evaporation	VWD	Isocratic; Eclipse XDB-C8 column (4.6 × 150 mm, 5 μm); mobile phase Methanol: Water (85: 15); flow rate 1 mL/ min	LOD: 0.004–0.038 mg/kg LOQ: 0.011–0.116 mg/kg	[48]
Pepper, chili pepper	Azo dyes: Dimethyl Yellow, Sudan I, II, III, IV, and Para Red	0.1% of formic acid in ACN → vigorous shaking → centrifugation → filtration	HR-Q-TOF-MS	Gradient; Acclaim 120 C18 column, (150 × 2.1 mm, 2.2 μm), mobile phase (A) 0.1% formic acid in 5 mM ammonium formate and (B) 0.1% formic acid in ACN; flow rate 0.3 ml/ml	Simple and rapid LOQ: 0.01 to 0.2 ng/g	[49]

Spices	Adulterant	Extraction process	Detection	LC condition	Selection criteria	Ref.
Beetroot red and paprika extract-based food colorant	Monascus red pigments (N-leucyl-rubropunctamine and its isomer N-isoleucyl-rubropunctamine)	Aqueous extraction by ultrasonication at room temperature → filtration	DAD, HR-Q-TOF-MS	Gradient; Kinetex C18 column (50 × 4.6 mm, 2.6 μm); mobile phase (A) ACN, (B) water (C) 1% formic acid; flow rate 0.8 ml/min	Excellent regulatory identification point	[50]
Chili products	Fourteen Fat soluble dyes including azo dyes	Ultrasound-assisted extraction with acetone-hexane mixture	DAD	Gradient; Agilent XDB C18 column (250 x 4.6 mm, 5 μm) at 30°C; mobile phase (A) 0.1% formic acid, (B) methanol: ACN 50:50, v/v; flow rate	Accurate and repeatable; GPC clean-up LOD: 11–71 μg/kg	[51]
Chili containing products	Rhodamine B	Vortexed with methanol → ultrasonication → filtration	FLD	Isocratic; Agilent XDB-C18 column (4.6 x 150 mm, 5 μm) at 35°C; mobile phase methanol: water (65: 35 v/v); flow rate 1.1 ml/min	Simple, rapid, and sensitive; LOD: 3.7 μg/kg, LOQ:10 μg/kg	[52]
Chili or pepper-containing products	Para Red, Sudan Orange G, Sudan I, II, III, IV, Sudan Red 7B, Rhodamine B, and Tropaeolin 000	Shaken with 1% acetic acid: acetonitrile (5: 95 v/v) → filtration	MS	Gradient; Purospher Star RP-18 endcapped column (125 × 3 mm; 5 μm) at 40°C; mobile phase (A) acetonitrile and (B) 5% acetic acid; flow rate 0.5 ml/min	Simple and rapid; LOD: 0.02–0.50 mg/kg	[53]

Abbreviations: ACN: acetonitrile, PDA: photodiode array, SPE: solid phase extraction, DAD: diode array detection, MS: mass detection, MS/MS: tandem mass spectrometry, VWD: variable wavelength detection, HR-Q-TOF: high-resolution quadrupole time-of-flight mass spectrometry, GPC: gel permeation chromatography, FLD: fluorescence detection.

Table 1.
List of adulterant dyes detected from spices by using HPLC analysis.

chromatography (HPLC) is a highly accurate and reliable method for identifying the presence of synthetic dyes in spices. To identify these dyes in spices using HPLC, a sample of the spice is first extracted with a suitable solvent, preferably methanol or ACN. The extract is then filtered and injected into the HPLC column. The column is typically equipped with a UV-VIS detector that can detect the characteristic absorption spectra of the synthetic dyes. Different synthetic dyes have distinct retention times and spectral properties, making them easy to identify using HPLC. For example, Sudan I, II, III, and IV are commonly used to enhance the color of chili powder, and their presence can be detected using HPLC with UV-VIS detection. These dyes have characteristic absorption spectra with peaks at 480 nm, 503 nm, 528 nm, and 440 nm, respectively. By comparing the retention times and spectral properties of the synthetic dyes in the sample to those of known standards, it is possible to identify and quantify the levels of synthetic dyes in the spice sample.

A study by Yun et al. developed an HPLC method for the simultaneous identification and quantification of six synthetic dyes in spice samples, including chili powder and paprika. The authors found that some of the samples were contaminated with Sudan I and II [38]. Another study by Duan et al. used HPLC-MS/MS to analyze 15 synthetic dyes in chili powder samples from China. They discovered seven of the samples were contaminated with illegal dyes such as Sudan I and Rhodamine B [39]. Adulteration of Sudan I and II and Rhodamine B was further confirmed by Maria et al. in chili powder and chili powder using an HPLC-MS/MS method [40].

The immense usability and popularity of HPLC method for detection of adulterants in spices are because of its sensitivity, accuracy, and low detection limits. For instance, Zhang et al. developed an HPLC method for the determination of 14 synthetic dyes in 16 spice and seasoning samples. The method was found to be reliable and sensitive, with detection limits ranging from 0.005 to 0.05 mg/kg. Similarly, Wang et al. used HPLC-MS/MS to determine the synthetic dyes and succeed in determining eight synthetic dyes in 30 spice samples [41]. The limit of detection was ranging from 0.005 to 0.05 mg/kg. Zhang et al. used HPLC-PDA (photodiode array) detector and successfully identified 16 synthetic dyes from 30 spices with a detection limit ranging from 0.003 to 0.02 mg/kg [42]. Hu et al. developed an HPLC-MS/MS method for the determination of seven synthetic dyes in various spices [43]. The method was found to be sensitive and accurate, with detection limits ranging from 0.003 to 0.05 mg/kg. Apart from this literature, **Table 1** summarizes various other dyes used as adulterants for various spices and their detection method. The extraction and detection methods of synthetic dyes are also listed in **Table 1**.

From the above illustration, it can be understood that HPLC is a highly effective method for identifying synthetic dyes in spices. The UV-VIS detection system or mass spectroscopy coupled with HPLC can further assure the accuracy of the experiment. Comparing the retention time of test sample with the standard sample in UV-VIS method or standard library searching for MS analysis is more helpful for identifying and quantifying synthetic dyes in the spices. This can further help to ensure that food products are safe and free from harmful additives.

3. Quantification of pesticides and herbicides in spices by HPLC

Pesticides and herbicides are often used in the cultivation of spices to protect the crops from pests and weeds [54]. However, these chemicals can pose a risk to human health if consumed in excessive amounts. HPLC is a powerful analytical technique that can be used to quantify the levels of pesticides and herbicides in spices [22, 55]. To quantify pesticides and herbicides in spices by HPLC, a sample of the spice is first extracted, filtered, and then injected into the HPLC column. The column is typically equipped with a UV-VIS detector, and the elute is monitored at a specific wavelength that corresponds to the absorption spectra of the target pesticides or herbicides. Different pesticides and herbicides have different retention times and spectral properties, making them easy to identify and quantify using HPLC. The retention time is influenced by the chemical and physical properties of the compound, such as its molecular weight, polarity, and solubility. Once the retention time of the target pesticide or herbicide is determined, a calibration curve is constructed by analyzing a series of standard solutions containing known amounts of the target compound. The calibration curve allows for the quantification of the levels of the target compound in the sample, based on the peak area or height of the eluate corresponding to the target compound. In some cases, it may be necessary to use a mass spectrometer in combination with HPLC to identify and quantify trace levels of pesticides and herbicides in spices. Mass spectrometry can also provide additional information on the molecular mass of the target compound, allowing for more accurate identification and quantification. In this section, extensive literature review of published articles in the last 5 years focused on the quantification of pesticides and herbicides in spices by HPLC is furnished.

A study by Tesemma et al. used HPLC to analyze the presence of different pesticides in lemon, black pepper, and fenugreek seed samples and the most commonly detected pesticide was chlorpyrifos present in the level of 1.6 to 1.9 μg/kg [56]. In a different study by Jiao et al., HPLC was used to quantify the levels of pesticides in black cumin samples [57]. The authors found that the most commonly detected pesticides were pyrethroids and that the levels of pesticides were higher in samples that had been stored for longer periods of time. Wei et al. also used QuEChERS coupled with HPLC-MS/MS and quantified clothianidin and acetamiprid in black pepper samples [58]. Ultra-HPLC-quadrupole-orbitrap mass spectrometry was used by Arnab et al. for the detection of pesticide content in various spices such as chili, coriander, black pepper, cardamom, turmeric, etc. and the detection limit was 2 to 5 ng/ml [59]. Xuan et al. developed aLC-Q-TOF/M method for the simultaneous detection of various pesticide residues in chili and Sichuan peppesamplesand the LOQ of ≤5 μg kg was detected [60]. They reported that some of the samples were contaminated with imidacloprid. An illegal pesticide (chlorpyrifos) was detected by Yep et al. using HPLC method to quantify the pesticide in black pepper samples [61]. The method was found to be accurate and reliable.

From the above discussion, it can be noted that the presence of pesticide residues in spices is a significant problem and HPLC is an effective method for the determination of pesticide residues in spices. This study also demonstrates the continuous and effective application of HPLC for the quantification of pesticides and herbicides in spices.

4. Detection of mycotoxins in spices by HPLC

Mycotoxins are toxic secondary metabolites produced by certain molds that can contaminate various food products, including spices [62, 63]. The presence of mycotoxins in spices can pose a significant risk to human health, as some mycotoxins are carcinogenic or can cause other adverse health effects [64]. HPLC has become an analytical tool for the detection and quantification of mycotoxins in spices [65].

Spices	Adulterant	Extraction process	Detection	LC condition	Selection criteria	Ref.
Composite spices (biryani, nihari, and korma masala)	Aflatoxin B1, B2, G1, G2	Blend with ACN: water (60:40) → filtration	FLD	Isocratic; LiChroCART® 100 Å RP-18, (250 × 4.0 mm, 5 μm); mobile phase: Water: ACN: Methanol (65: 17.5: 17.5 v/v/v); flow rate: 1 ml/min	Rapid, sensitive, and accurate; LOD: 0.03–0.06 μg/kg LOQ: 0.09–0.30 μg/kg	[75]
Clove, black and white peppers, rosemary, and fennel	Mycotoxins	Blend with ACN: water (84:16) → filtration	FLD	Isocratic; C18 monolithic columns (100 × 4.6 mm); mobile phase 0.2% acetic acid: acetonitrile: methanol (40: 30: 30, v/v/v); flow rate: 1 mL/min	Not specified	[76]
Sesame seeds and products	Aflatoxin B1	Shaken with methanol: water (80: 20) → filtration	FLD	Isocratic; C18Nova-Pak (4.6 × 250 mm, 60 Å, 4 μm); mobile phase water: ACN: methanol (20:4:3); flow rate 1 ml/min	Sensitive; LOD: 0.02 ng/ml LOQ: 0.23 ng/ml	[77]
Mixed spices samples	Aflatoxins B1, B2, G1, and G2	extracted with methanol: water (80: 20 v/v) → purified → SPE	FLD	Isocratic: LiChroCART® 100 Å RP-18, (250 × 4.0 mm 5 μm,); mobile phase methanol: CAN: buffer (17.5: 17.5: 65 v/v/v) flow rate 1 mL/min	Rapid and sensitive LOD: 0.020–0.026 μg/kg LOQ: 0.062– 0.080 μg/kg	[78]
Chili pepper	Peanuts allergens	Homogenize with buffer(30% sucrose, 0.1 M KCl, 50 mM Tris-HCl pH 8, and 5 mM EDTA) → Centrifused	MS/MS	Gradient; AcquityTM UPLC BEH300 C18 RP column (2.1 × 150 mm 300 Å, 1.7 μm); mobile phase 1 ml of 2% ACN, 0.1% TFA, and 2 ml of 70% ACN +0.1% TFA, flow rate 1 mL/min	LOD: 24 mg/kg	[79]
Cinnamon	C. Cassia	Sonicate with methanol → centrifuge	UV/MS	Gradient; Acquity UPLC BEH shield RP18 column (100 mm × 2.1 mm, 1.7 μm); mobile phase water with 0.05% formic acid (A) and methanol/ ACN (90:10, v/v) with 0.05% formic acid (B), flow rate 0.23 mL/min	LOQ: o.2 μg/kg	[80]
Black pepper, basil, oregano, nutmeg, paprika, and thyme	Mycotoxins, pesticides, and toxic metals	SPE	MS/MS	Isocratic; Kinetex C18 analytical column(2.1 × 2.6 μm); mobile phase aqueous 1% formic acid and 100% methanol, flow rate 0.3 mL/ min	LOD: 0.8 μg/g LOQ: 2.7 μg/g	[54]

Spices	Adulterant	Extraction process	Detection	LC condition	Selection criteria	Ref.
Smoked chili pepper	Aflatoxins (B1, B2, G1, and G2)	Not available	FLD		Good separation LOD: 1–20 ng/g	[72]
Paprika and chili	Aflatoxin B1, B2, G1, G2, ochratoxin A, mycotoxin	Blend with methanol: water (8:2) → filtration	FID	Isocratic; ODS2 (4.6 × 150 mm) column; mobile phase water: CAN: methanol (60:25:15 v/v/v); flow rate 1 ml/min	LOD ranges from 034 to 4.77 μg/Kg	[81]
Black pepper, turmeric, nutmeg, etc.	Aflatoxin B1, G1	Extracted with methanol and water (80:20)	FLD	Gradient; RP-C18 column; mobile phase water: ACN: methanol (60:20:20) flow rate	LOD: 0.2–0.5 μg/kg LOQ: 0.6–1.5 μg/kg	[82]
Garlic, ginger, pepper, etc.	Aflatoxin B1, B2, G1 and G2	Extracted with methanol, deionized water, hexane, and NaCl	FLD	Isocratic; C18 column; (4.6 × 250 mm × 5 μm); mobile phase methanol: deionized water (80:20), Flow rate 1 ml/min	LOD: 0.24– 8.56 μg/kg and LOD: 0.11–3.68 μg/kg	[83]
Cinnamon, clove, cardamom, and ginger	Aflatoxin B1, B2, G1, and G2	Extracted with water: methanol: ACN (60:30:10, v/v/v)	FLD	Eclipse XDB- C18 (5 μm, 4.6 × 150 mm) column; flow rate of 1.2 ml/min	Simple and rapid multi-residue method LOD: 0.13– 0.16 μg/kg; LOQ: 0.16–0.29 μg/kg	[84]
Mixed spices	Citrinin	Blended with methanol → filtration	FLD	Isocratic; Hypercil GOLD LC column (150 × 4.6 mm, 3 μm); mobile phase ACN: 10 mM phosphoric acid pH 2.5 (50: 50 v/v); flow rate 1 mL/min.	Reliable and sensitive cleanup; LOD: 1 μg/kg; LOQ: 3 μg/kg;	[85]
White and black pepper	Aflatoxins	Extracted with ACN: methanol: water (1:3:6)	FLD	Isocratic; ODS (250 × 4.6 mm i.d., 5 μg); Flow rate of 0.5 mL/min	Simple and rapid multi-residue method LOD:m 0.06–0.11 (μg/ml) LOQ:0.21–0.36 (μg/ml)	[86]
Multiple dried spices	Aflatoxin B1	Stirring with methanol: Water (80: 20 v/v) → filtration → SPE	FLD	Isocratic; C18 Nova-Pak (4.6 × 250 mm, 4 μm); mobile phase water: acetonitrile: methanol (20:4:6 v/v/v); flow rate 1 ml/min	Good recovery LOD: 0.1 ng/g LOQ: 0.45 ng/g	[74]

Spices	Adulterant	Extraction process	Detection	LC condition	Selection criteria	Ref.
Dried and mixed spices (chili, fennel, cumin, turmeric, black and white pepper, poppy seed, and coriander)	Aflatoxins (B1, B2, G1 and G2) and ochratoxin A	Sample with sodium chloride → 1st: extracted with methanol: water (7: 3, v/v); 2nd: methanol–water (8:2, v/v) → shaken	FLD	Isocratic; CLC-ODS column (250 × 4.6 mm, 5 μm) at 40°C; mobile phase water: methanol: acetonitrile (70:20:10 v/v/v); flow rate 1 ml/ min	High specificity; LOD: Aflatoxin, 0.01 ng/g ochratoxin A, 0.10 ng/g LOQ: Aflatoxin, 0.6–1.2 ng/g ochratoxin A, 2.1–8.0 ng/g	[71]

Table 2.
Recent reporting on aflatoxin detection in spices through HPLC method.

This section summarizes some of the recent findings where HPLC is used for the detection of mycotoxins in various spices.

A study reported by Iqbal et al. showed simultaneous detection of aflatoxins and ochratoxin A in spices by HPLC method. The developed HPLC method was found to be reliable and sensitive and was able to detect mycotoxins at concentrations as low as 0.5 μg/kg [66]. Da Silva et al. also used HPLC to analyze the presence of Ochratoxin A in black pepper (29 powder and 31 grains) and found the range of this mycotoxin between 0.05 and 13.15 μg/kg [67]. Zareshahrabadi et al. reported that out of 80 spice samples, 40 were contaminated with aflatoxins and 48 were ochratoxin A (some were common spices where both the mycotoxins were identified) through HPLC [68]. Ainiza et al. checked aflatoxins in fennel, coriander, turmeric, cumin, and chili and found and reported aflatoxins B1, B2, G1, and G2 through HPLC. The authors further stated that the optimized HPLC method can be utilized for the detection of various other mycotoxins from adulterated spices [69]. Aflatoxins (AFs) and ochratoxin A (OTA) contamination in *C. annuum* was reported by Costa et al. [70]. In a different report from Ali et al., aflatoxins and ochratoxin A detection were carried out by HPLC coupled with fluorescent detector on dried chili, black and white pepper, coriander, turmeric, fennel, cumin, poppy seed, etc. The limit of detection for aflatoxin was found to be 0.01 ng/g and 0.10 ng/g for ochratoxin A [71]. Palma et al. optimized and validated the quantification of various aflatoxin in Marken using HPLC-FLD (fluorescence detection). The optimized method furnished LOD with a minimum range of 0.6–20 ng/g [72]. Aflatoxin adulteration in coriander seed was checked by Ouakhssase et al. using UPLC MS/MS. The limit of quantification was found to be between 0.12 to 0.5 μg/kg [73]. This method validated aflatoxin contamination in coriander seeds successfully. Koutsias et al. confirmed presence of aflatoxins B1 in various spices of Greece [74]. HPLC with fluorescence detector was used for the detection of aflatoxins and the quantification limits were ascertained at 0.1 ng/g and 0.45 ng/g.

Mycotoxins are the unwanted toxic toxin released by the pathogens formed in spices, which are poorly stored or processed. Their detection is very important for the quality control of the spices. Therefore, employment of HPLC for the detection of mycotoxins in spices is considered vastly. The aforementioned literature review suggested that HPLC coupled with UV-VIS detector or mass spectroscopy has successfully identified many mycotoxins in wised varieties of spice samples. Some of the recent reporting on different aflatoxin-contaminated spices and their detection through HPLC is presented in **Table 2**.

5. HPLC analysis of essential oils in spices

Essential oils are natural products that are extracted from various parts of plants, including leaves, stems, flowers, and fruits [87, 88]. They are widely used in the food and beverage, cosmetics, and pharmaceutical industries for their flavor, aroma, and medicinal properties [89]. However, substandard drugs are used for adulterating the genuine crude drugs. For instance, de fat clove, funnel, and coriander are mixed with their genuine counterparts. HPLC finds its way to the analysis of essential oil composition in substandard spices [90]. There are reports suggesting coupling of HPLC with UV-VIS detector or MS offers more accuracy for the detection of essential oil [91].

Bendif et al. used HPLC to analyze the essential oil content and chemical composition of thyme. The study found that the major component of the essential oils in thyme was monoterpenes and sesquiterpenes and oxygenated monoterpenes [92].

The authors suggested that HPLC can be an effective tool for the identification and quantification of the major components of essential oils in spices. Chen et al. also used HPLC to analyze the essential oil content and composition of black pepper and white pepper to report higher concentrations of monoterpenes and sesquiterpenes [93]. A study by Ling et al. reported the use of HPLC-MS to identify and quantify essential oil including cinnamaldehyde in cinnamon bark [94]. In a study by Ji et al., HPLC was used to analyze the essential oil components in Sichuan pepper and confirmed the presence of monoterpenes [95]. HPLC is a method of choice for analyzing less volatile or nonvolatile constituents in essential oils. HPLC detects nonvolatile adulteration, such as synthetic compounds or vegetable oils [96, 97]. Ding et al. used HPLC-based fingerprint analysis to evaluate quality of 24 cinnamon bark and 32 cinnamon twig samples sourced from various countries. The study separated and determined seven major marker compounds: cinnamaldehyde, eugenol, coumarin, cinnamyl alcohol, cinnamic acid, 2-hydroxyl cinnamaldehyde, and 2-methoxy cinnamaldehyde [98]. Lee determined the total volatile material contents of black pepper and white pepper via the SDE (Linkens-Nikens type simultaneous steam distillation and extraction apparatus) aided HPLC method [99]. In addition, Yeh et al. determined the essential oil content in two different varieties of ginger root using HPLC analysis [100]. A study by Yang et al. reported use of HPLC-MS/MS to identify 101 small molecules including flavonols and flavones, phenolic acids, lactones, terpenoids, phenylpropanoids, and flavanols from waste cinnamon leaves [94].

This section summarizes the importance of HPLC in determining various essential oil in spices. By the application of HPLC method, substandard varieties of spices can be avoided for commercial marketing. HPLC analysis for essential oil determination can also limit chances of spice adulteration.

6. HPLC quantification of starch and other fillers in spices

The addition of fillers to spices is a common form of adulteration, which can lead to reduced quality, taste, and nutritional value of the spice [101]. One of the most common fillers used in spices is starch, which can be derived from various sources, such as corn, wheat, or rice [102]. Starch quantification is carried out by first extracting the sample with water or ethanol followed by filtration and subjection into the HPLC column. The column is typically equipped with a refractive index detector (RID), which detects changes in the refractive index of the eluent as the sample components pass through the column [103, 104]. Excessive fillers destroy the authenticity of the spices, and to detect the fillers in the powdered spices, HPLC is an alternate solution. In this section, some of the recent findings on detection of starch and fillers in spices have been mentioned.

Ordoudi et al. [105] controlled saffron shelf life using FT-MIR spectra. They confirmed presence of glucose molecules and glycoside linkage damage in spice samples at 1028 cm^{-1} and in the range 1175–1157 cm^{-1} band intensity, respectively. The obtained FT-MIR spectroscopic data were analyzed using PCA. To verify the outcomes of the FT-MIR-PCA procedure, HPLC-DAD analysis was carried out [106]. HPLC-DAD method was used to check the authenticity of the results obtained from FT-MIR-PCA analysis. This signifies that HPLC method for detection of unwanted and harmful adulterants is the most reliable method of choice.

Starch and other fillers can be added to spices to increase their weight and bulk, making them more profitable for manufacturers. However, the presence of fillers in spices can reduce their quality and potentially cause health hazards. Therefore, it is important to determine the amount of fillers in spices to ensure that they meet the standards for purity and quality.

7. HPLC in comparison to other analytical techniques in quality control of spices

This section is based on the comparison of HPLC with other physical, chromatographical, and spectroscopical methods of adulteration detection in spices. Several analytical techniques are in use for checking the authenticity of spices. These methods include physical authentication techniques and spectroscopic and chromatographic analysis. Macroscopic and microscopic standardization of spices are done under the physical authentication method. Color, flavor, shape, size, etc. are examined personally by workers to identify the adulteration in the macroscopic analysis method. On the other hand, arrangement of tissue, tissue layering, fiber structure, and root and rhizome structure is checked through microscopic lens in the microscopic method. These methods have become outdated because they are nonreliable, lack repeatability, and time-consuming. These methods also require many skilled people for the microscopic studies. Moreover, different spectroscopic methods such as FTIR, NMR, Raman, XRD, and mass spectroscopy have garnered popularity for detection of illegal substances in spices. They analyze the ingredients and structure of spices efficiently and are very sensitive and accurate techniques. However, they are very expensive and also come up with very complex software algorithm to analyze the results.

Other chromatographic techniques such as gas chromatography (GC) and thin-layer chromatography (TLC) are the commonly used separation methods for the isolation and detection of adulterants in spices. However, each method has its advantages and limitations. GC is widely used for the detection of mycotoxins in spices, but it requires derivatization and has limited selectivity for nonvolatile compounds. TLC is a cost-effective and rapid method, but it lacks sensitivity and requires extensive sample preparation.

Spectrophotometry is a simple and rapid method, but it lacks specificity and is limited to certain compounds. Infrared spectroscopy is a nondestructive method, but it is less sensitive and requires skilled operators. In contrast, HPLC offers high sensitivity, specificity, and selectivity for a wide range of compounds, including mycotoxins, synthetic dyes, pesticides, herbicides, and starch. Moore et al. [17] found that when comparing different methods for identifying compounds and adulterants, IR spectroscopy had the second-highest number of references after HPLC paired with a specific type of detection equipment. Nevertheless, mass spectrometry [107] and IR spectroscopy techniques were found to have the most reports of adulterants in the literature.

HPLC has several advantages over other analytical methods for the detection of adulterants in spices. Firstly, it can separate and identify multiple compounds in a single run, which saves time and reduces the cost of analysis. Secondly, HPLC is highly sensitive and can detect adulterants at low levels, ensuring the safety and quality of spices. Thirdly, HPLC is highly specific and can distinguish between different compounds, ensuring the accuracy and reliability of results. However, there are some limitations associated with HPLC that require expensive equipment and skilled operators, which may limit its use in some settings. Further HPLC may require extensive sample preparation, which may increase the cost and time of analysis.

Finally, HPLC may not be able to detect certain compounds that are not amenable to its separation and detection methods.

8. Conclusions

Now days food quality and safety have become one of major concerns because of the serious health hazards observed due to a range of food intake. However, various fraudulent activities that include adulteration of toxic material into spices have become a serious concern. To address this serious issue, different physical, microscopical, and analytical methods have been carried out. HPLC is one of the chromatographic techniques relied on mostly for the detection of adulteration in spices. The combination of HPLC with other analytical instruments such as mass spectrometry, NMR, etc. is found to be efficient in identifying adulterants in trace amounts also. This method has become most trustworthy because of its sensitivity, accuracy, and reproducibility. The current chapter explains the recent application of HPLC for the quality control of spices. HPLC coupled with UV-VIS or MS/MS detectors have been proven effective in determination of mycotoxins, starch, fillers, and pesticides in verity of spices. Different carcinogenic and synthetic dyes are also successfully identified through HPLC analysis. This present study furnishes extensive scientific information related to utilization of HPLC for the detection of various harmful adulterants in spices. However, more literature on detection of powder fillers and spoiled volatile oils in spices (clove, coriander, and fennel) by HPLC analysis is due in this study.

Conflict of interest

The authors declare no conflict of interest for this publication.

Author details

Prafulla Kumar Sahu*, Sukumar Purohit, Swarnajeet Tripathy, Durga Prasad Mishra and Biswajeet Acharya
School of Pharmacy, Centurion University of Technology and Management, Odisha, India

*Address all correspondence to: prafulla.sahu@cutm.ac.in

References

[1] Sasikumar B et al. Advances in adulteration and authenticity testing of herbs and spices. In: Advances in Food Authenticity Testing. Woodhead Publishing: Elsevier; 2016. pp. 585-624

[2] Jessica Elizabeth DLT et al. Spice use in food: Properties and benefits. Critical Reviews in Food Science and Nutrition. 2017;**57**:1078-1088

[3] Martínez-Graciá C et al. Use of herbs and spices for food preservation: Advantages and limitations. Current Opinion in Food Science. 2015;**6**:38-43

[4] Nair KP, The Agronomy and Economy of Turmeric and Ginger: The Invaluable Medicinal Spice Crops: Newnes. London: Elsevier; 2013

[5] Sanlier N, Gencer F. Role of spices in the treatment of diabetes mellitus: A minireview. Trends in Food Science & Technology. 2020;**99**:441-449

[6] M. Manual, Web Directory for Organic Spices, Herbs and Essential Oils, 2006.

[7] Sudhabindu K, Samal L. Common adulteration in spices and Do-at-home tests to ensure the purity of spices. Food Science Repoets. 2020;**1**:66-68

[8] Sicherer SH, Sampson HA. 9. Food allergy. Journal of Allergy and Clinical Immunology. 2006;**117**:S470-S475

[9] Nallappan K et al. Identification of adulterants in turmeric powder using terahertz spectroscopy. In: 2013 38th International Conference on Infrared, Millimeter, and Terahertz Waves (IRMMW-THz). Germany; 2013. pp. 1-2

[10] WHO. Exposure to highly hazardous pesticides: A major public health concern. 2010

[11] Spink J, Moyer DC. Defining the public health threat of food fraud. Journal of Food Science. 2011;**76**:R157-R163

[12] Tarantelli T, Sheridan R. Toxic Industrial Colorants Found in Imported Foods. New York: State Department of Agriculture & Markets Food Laboratory; 2011

[13] Galvin-King P et al. Herb and spice fraud; the drivers, challenges and detection. Food Control. 2018;**88**:85-97

[14] Moses J et al. Climate change and its implications on stored food grains. Agricultural Research. 2015;**4**:21-30

[15] Ismaiel AA, Papenbrock J. Mycotoxins: Producing fungi and mechanisms of phytotoxicity. Agriculture. 2015;**5**:492-537

[16] Kabak B, Dobson AD. Mycotoxins in spices and herbs–an update. Critical Reviews in Food Science and Nutrition. 2017;**57**:18-34

[17] Moore JC et al. Development and application of a database of food ingredient fraud and economically motivated adulteration from 1980 to 2010. Journal of Food Science. 2012;**77**:R118-R126

[18] Babaei S et al. Developing an SCAR and ITS reliable multiplex PCR-based assay for safflower adulterant detection in saffron samples. Food Control. 2014;**35**:323-328

[19] Reemtsma T. Liquid chromatography–mass spectrometry and strategies for trace-level analysis of polar organic pollutants. Journal of Chromatography A. 2003;**1000**:477-501

[20] Dalby A. Dangerous Tastes, The Story of Spices. Berkeley: University of California Press; 2000

[21] Freedman P. Out of the East: Spices and the Medieval Imagination. USA: Yale University Press, Chicago press; 2008

[22] Rodinson M, Arberry AJ, "Medieval Arab Cookery". USA: Prospect books; 2001

[23] Galli M. Beyond frontiers: Ancient Rome and the Eurasian trade networks. Journal of Eurasian Studies. 2017;**8**:3-9

[24] Bharathi S et al. Instrument-based detection methods for adulteration in spice and spice products–A review. Journal of Spices and Aromatic Crops. 2018;**27**:106-118

[25] Kar M. Towards sustainable indian Agri-commodities' markets: Experiences, innovative model and future agenda. Pragati: Journal of Indian Economy. 2020;**7**:48-63

[26] Bansal S et al. Food adulteration: Sources, health risks, and detection methods. Critical Reviews in Food Science and Nutrition. 2017;**57**: 1174-1189

[27] Jaffee S. Delivering and Taking the Heat: Indian Spices and Evolving Product and Process Standards. USA: The World Bank group; 2005

[28] Kumari C, Singh SG. Mycoflora of spices, A review. Plant Archives. 2021;**21**:99-102

[29] DeWaal CS, Brito GRG. Safe food international: A blueprint for better global food safety. Food and Drug Law Journal. 2005;**60**:393-405

[30] Thomas L, Sanil P. Competitiveness in spice export trade from India: A review. Food Science and Biotechnology. 2019;**28**

[31] Moges A et al. Dietary and bioactive properties of the berries and leaves from the underutilized Hippophae salicifolia D. Don grown in Northeast India. Food Science and Biotechnology. 2021;**30**:1555-1569

[32] Purohit S et al. Exploration of nutritional, antioxidant and antibacterial properties of unutilized rind and seed of passion fruit from Northeast India. Journal of Food Measurement and Characterization. 2021;**15**:3153-3167

[33] Sahu PK et al. An overview of experimental designs in HPLC method development and validation. Journal of Pharmaceutical and Biomedical Analysis. 2018;**147**:590-611

[34] Bhooma V et al. Identification of synthetic dyes magenta III (new fuchsin) and rhodamine B as common adulterants in commercial saffron. Food Chemistry. 2020;**309**:125793

[35] Sahu PK et al. A robust RP-HPLC method for determination of turmeric adulteration. Journal of Liquid Chromatography & Related Technologies. 2020;**43**:247-254

[36] Vickers NJ. Animal communication: When i'm calling you, will you answer too? Current Biology. 2017;**27**:R713-R715

[37] Cokosyler N. Farkh yontemlerle kurutulan kirmizi biberlerde Aspergillus flavus gelisimi ve aflatoksin olusumunun incelenmesi. Gida. 1999;**24**:297-306

[38] Zhu Y et al. A screening method of oil-soluble synthetic dyes in chilli products based on multi-wavelength chromatographic fingerprints comparison. Food Chemistry. 2016;**192**:441-451

[39] Duan H-L et al. Magnetically modified porous β-cyclodextrin polymers for dispersive solid-phase extraction high-performance liquid chromatography analysis of Sudan dyes. Food Analytical Methods. 2019;**12**:1429-1438

[40] Khalikova MA et al. On-line SPE–UHPLC method using fused core columns for extraction and separation of nine illegal dyes in chilli-containing spices. Talanta. 2014;**130**:433-441

[41] He T et al. Dummy molecularly imprinted polymer based microplate chemiluminescence sensor for one-step detection of Sudan dyes in egg. Food Chemistry. 2019;**288**:347-353

[42] Zhang M et al. A composite polymer of polystyrene coated with poly (4-vinylpyridine) as a sorbent for the extraction of synthetic dyes from foodstuffs. Analytical Methods. 2020;**12**:3156-3163

[43] Hu Z et al. Simultaneous determination of multiclass illegal dyes with different acidic–basic properties in foodstuffs by LC-MS/MS via polarity switching mode. Food Chemistry. 2020;**309**:125745

[44] Cornet V et al. Journal of Agricultural and Food Chemistry. 2006;**54**:639-644

[45] Genualdi S et al. Method development and survey of Sudan I–IV in palm oil and chilli spices in the Washington, DC, area. Food Additives & Contaminants: Part A. 2016;**33**:583-591

[46] Sebaei AS et al. Determination of seven illegal dyes in Egyptian spices by HPLC with gel permeation chromatography clean up. Journal of Food Composition and Analysis. 2019;**84**:103304

[47] Bessaire T et al. A new high-throughput screening method to determine multiple dyes in herbs and spices. Food Additives & Contaminants: Part A. 2019;**36**:836-850

[48] Ullah A et al. Banned Sudan dyes in spices available at markets in Karachi, Pakistan. Food Additives & Contaminants: Part B. 2022;**2022**:1-8

[49] Amelin V et al. Simultaneous determination of dyes of different classes in aquaculture products and spices using HPLC–high-resolution quadrupole time-of-flight mass spectrometry. Journal of Analytical Chemistry. 2017;**72**:183-190

[50] Thalhamer B, Buchberger W. Adulteration of beetroot red and paprika extract based food colorant with Monascus red pigments and their detection by HPLC-QTof MS analyses. Food Control. 2019;**105**:58-63

[51] Zhu Y et al. Simultaneous determination of 14 oil-soluble synthetic dyes in chilli products by high performance liquid chromatography with a gel permeation chromatography clean-up procedure. Food Chemistry. 2014;**145**:956-962

[52] Qi P et al. Development of a rapid, simple and sensitive HPLC-FLD method for determination of rhodamine B in chili-containing products. Food Chemistry. 2014;**164**:98-103

[53] Botek P et al. Determination of banned dyes in spices by liquid chromatography-mass spectrometry. Czech Journal of Food Science. 2007;**25**:17-24

[54] Reinholds I et al. Mycotoxins, pesticides and toxic metals in commercial spices and herbs. Food Additives & Contaminants: Part B. 2017;**10**:5-14

[55] Wang Y et al. Analytical methods to analyze pesticides and herbicides.

Water Environment Research. 2019;**91**:1009-1024

[56] Mekonnen TF et al. Investigation of chlorpyrifos and its transformation products in fruits and spices by combining electrochemistry and liquid chromatography coupled to tandem mass spectrometry. Food Analytical Methods. 2018;**11**:2657-2665

[57] Ersoy N et al. Determination of pesticide residue present in cumin plant (Nigella orientalis L.) with LC-MS/MS and GC-MS. Asian Journal of Chemistry. 2016;**28**:1011

[58] Yao W et al. Multi-residue analysis of 34 pesticides in black pepper by QuEChERS with d-SPE vs. d-SLE cleanup. Food Analytical Methods. 2019;**12**:176-189

[59] Goon A et al. A simultaneous screening and quantitative method for the multiresidue analysis of pesticides in spices using ultra-high performance liquid chromatography-high resolution (Orbitrap) mass spectrometry. Journal of Chromatography A. 2018;**1532**:105-111

[60] Liu X et al. Determination of pesticide residues in chilli and Sichuan pepper by high performance liquid chromatography quadrupole time-of-flight mass spectrometry. Food Chemistry. 2022;**387**:132915

[61] Yap C, Jarroop Z. Residue levels and dissipation behaviors of chlorpyrifos in black pepper berries and soil. Food Research. 2018;**2**:587-593

[62] Milićević DR et al. Real and perceived risks for mycotoxin contamination in foods and feeds: Challenges for food safety control. Toxins. 2010;**2**:572-592

[63] Tahir NI et al. Nature of aflatoxins: Their extraction, analysis, and control. Journal of Food Safety. 2018;**38**:e12561

[64] Bryden WL. Mycotoxins in the food chain: Human health implications. Asia Pacific Journal of Clinical Nutrition. 2007;**16**:95-101

[65] Rahmani A et al. Qualitative and quantitative analysis of mycotoxins. Comprehensive Reviews in Food Science and Food Safety. 2009;**8**:202-251

[66] Iqbal SZ et al. Survey of aflatoxins and ochratoxin A in retail market chilies and chili sauce samples. Food Control. 2017;**81**:218-223

[67] da Silva ARP et al. Ochratoxin A and related fungi in Brazilian black pepper (Piper nigrum L.). Food Research International. 2021;**142**:110207

[68] Zareshahrabadi Z et al. Detection of aflatoxin and ochratoxin A in spices by high-performance liquid chromatography. Journal of Food Quality. 2020;**2020**:1-8

[69] Ainiza WW et al. Simultaneous determination of aflatoxins and ochratoxin A in single and mixed spices. Food Control. 2015;**50**:913-918

[70] Costa J et al. Occurrence of aflatoxins and Ochratoxin A during Merkén pepper powder production in Chile. Food. 2022;**11**:3843

[71] Ali N et al. Natural occurrence of aflatoxins and ochratoxin A in processed spices marketed in Malaysia. Food Additives & Contaminants: Part A. 2015;**32**:518-532

[72] Palma P et al. Adaptation, optimization, and validation of a sensitive and robust method for the quantification of total aflatoxins (B1, B2, G1, and G2) in the spice merkén by HPLC-FLD with post-column derivatization. Microchemical Journal. 2022;**178**:107342

[73] Ouakhssase A et al. Modified-QuEChERS-LC/MS method for the analysis of aflatoxins and ochratoxin A in coriander seeds. Food Additives & Contaminants: Part A. 2023;**2023**:1-9

[74] Koutsias I et al. Occurrence and risk assessment of aflatoxin b1 in spices marketed in Greece. Analytical Letters. 2021;**54**:1995-2008

[75] Asghar MA et al. Aflatoxins in composite spices collected from local markets of Karachi, Pakistan. Food Additives & Contaminants: Part B. 2016;**9**:113-119

[76] Garcia MV et al. Aflatoxigenic and ochratoxigenic fungi and their mycotoxins in spices marketed in Brazil. Food Research International. 2018;**106**:136-140

[77] Kollia E et al. Aflatoxin B1 in sesame seeds and sesame products from the Greek market. Food Additives & Contaminants: Part B. 2016;**9**:217-222

[78] Asghar MA et al. Development and validation of a high-performance liquid chromatography method with post-column derivatization for the detection of aflatoxins in cereals and grains. Toxicology and Industrial Health. 2016;**32**:1122-1134

[79] Vandekerckhove M et al. Development of an LC-MS/MS method for the detection of traces of peanut allergens in chili pepper. Analytical and Bioanalytical Chemistry. 2017;**409**:5201-5207

[80] Wang Y-H et al. Cassia cinnamon as a source of coumarin in cinnamon-flavored food and food supplements in the United States. Journal of Agricultural and Food Chemistry. 2013;**61**:4470-4476

[81] Santos L et al. Co-occurrence of aflatoxins, ochratoxin A and zearalenone in Capsicum powder samples available on the Spanish market. Food Chemistry. 2010;**122**:826-830

[82] Romagnoli B et al. Aflatoxins in spices, aromatic herbs, herb-teas and medicinal plants marketed in Italy. Food Control. 2007;**18**:697-701

[83] Akpo-Djènontin DOO et al. Mold infestation and aflatoxins production in traditionally processed spices and aromatic herbs powder mostly used in West Africa. Food Science & Nutrition. 2018;**6**:541-548

[84] Fundikira S et al. Awareness, handling and storage factors associated with aflatoxin contamination in spices marketed in Dar es Salaam, Tanzania. World Mycotoxin Journal. 2021;**14**:191-200

[85] Mair C et al. Assessment of Citrinin in spices and infant cereals using Immunoaffinity column clean-up with HPLC-fluorescence detection. Toxins. 2021;**13**:715

[86] Saito K et al. Residual analysis of aflatoxins in spice by HPLC coupled with solid-phase dispersive extraction and solid-phase fluorescence derivatization method. Journal of AOAC International. 2020;**103**:1521-1527

[87] Irshad M et al. "Biological importance of essential oils," Essential Oils-Oils of Nature. Vol. 1. London: IntechOpen; 2020

[88] Vankar PS. Essential oils and fragrances from natural sources. Resonance. 2004;**9**:30-41

[89] Irshad M et al. Biological importance of essential oils. Essential Oils-Oils of Nature. 2020;**1**

[90] Zhang J et al. Basic sensory properties of essential oils from aromatic plants and their applications: A critical

review. Critical Reviews in Food Science and Nutrition. 2023;**2023**:1-14

[91] Turek C, Stintzing FC. Application of high-performance liquid chromatography diode array detection and mass spectrometry to the analysis of characteristic compounds in various essential oils. Analytical and Bioanalytical Chemistry. 2011;**400**:3109-3123

[92] Bendif H et al. Total phytochemical analysis of Thymus munbyanus subsp. coloratus from Algeria by HS-SPME-GC-MS, NMR and HPLC-MSn studies. Journal of Pharmaceutical and Biomedical Analysis. 2020;**186**:113330

[93] Chen S-X et al. Comparison of chemical compositions of the Pepper EOs from different cultivars and their AChE inhibitory activity. Natural Product Communications. 2020;**15**:19345

[94] Yang Y-L et al. LC-MS-based identification and antioxidant evaluation of small molecules from the cinnamon oil extraction waste. Food Chemistry. 2022;**366**:130576

[95] Ji Y et al. Chemical composition, sensory properties and application of Sichuan pepper (Zanthoxylum genus). Food Science and Human Wellness. 2019;**8**:115-125

[96] Do TKT et al. Authenticity of essential oils. TrAC Trends in Analytical Chemistry. 2015;**66**:146-157

[97] Smelcerovic A et al. Recent advances in analysis of essential oils. Current Analytical Chemistry. 2013;**9**:61-70

[98] Ding Y et al. Discrimination of cinnamon bark and cinnamon twig samples sourced from various countries using HPLC-based fingerprint analysis. Food Chemistry. 2011;**127**:755-760

[99] Lee M-S. Simple rapid quality estimation method in black and white pepper grounds by determination of volatile oil content. The Korean Journal of Food and Nutrition. 2009;**22**:352-356

[100] Yeh H-y et al. Bioactive components analysis of two various gingers (Zingiber officinale Roscoe) and antioxidant effect of ginger extracts. LWT-Food Science and Technology. 2014;**55**:329-334

[101] Momtaz M et al. Mechanisms and health aspects of food adulteration: A comprehensive review. Food. 2023;**12**:199

[102] Adigwe OP et al. "Starch: A Veritable Natural Polymer for Economic Revolution," 2022.

[103] Murillo MMS, Granados-Chinchilla F. Total starch in animal feeds and silages based on the chromatographic determination of glucose. MethodsX. 2018;**5**:83-89

[104] Sluiter JB et al. Direct determination of cellulosic glucan content in starch-containing samples. Cellulose. 2021;**28**:1989-2002

[105] Ordoudi SA et al. On the quality control of traded saffron by means of transmission Fourier-transform mid-infrared (FT-MIR) spectroscopy and chemometrics. Food Chemistry. 2014;**150**:414-421

[106] Kucharska-Ambrożej K, Karpinska J. The application of spectroscopic techniques in combination with chemometrics for detection adulteration of some herbs and spices. Microchemical Journal. 2020;**153**:104278

[107] Osman AG et al. Overview of analytical tools for the identification of adulterants in commonly traded herbs and spices. Journal of AOAC International. 2019;**102**:376-385